Water quality and raw sewage spills!

By

Patricia E.A. Mallenby, BA, BSc

and

Jeremy T.T. Mallenby, BA, BSc

Preface

The authors recall the line, "Water, water … everywhere but not a drop to (safely) drink", has been adlibbed from *The Rime of the Ancient Mariner*[1]:

As cited, "the Rime of the Ancient Mariner (originally The Rime of the Ancyent Marinere) is the longest major poem by the English poet Samuel Taylor Coleridge, written in 1797–98 and was published in 1798 in the first edition of Lyrical Ballads."[2]

"Day after day, day after day,

We stuck, nor breath nor motion;

As idle as a painted ship

Upon a painted ocean.

Water, water, everywhere,

And all the boards did shrink;

Water, water, everywhere,

Nor any drop to drink."[3]

This book, in addition to using a simple "Urban Water Test" kit to satisfy their curiosity as to how safe was their tap water in Winnipeg, Manitoba, Canada?

They were also curious as to how the recent raw sewage spill affected the water quality [see Appendix i & ii, cited below]?

For example, as cited by the CBC News, "More than one billion litres of partially treated sewage have flowed into the Red River over the past few weeks, due to a major glitch at a Winnipeg waste treatment plant, according to city officials."[4]

"For an unknown reason, these microorganisms suddenly stopped thriving on Oct. 7, 2011, upsetting the full biological treatment process," the city stated in a release.[5]

Footnote

1 - 3. *The Rime of the Ancient Mariner*
http://webcache.googleusercontent.com/search?q=cache:vclFugZ9u-
cJ:en.wikipedia.org/wiki/The_Rime_of_the_Ancient_Mariner+Water,+water+%E2
%80%A6+everywhere+but+not+a+drop+to+(safely)+drink&cd=1&hl=en&ct=clnk
&gl=ca

4 - 5. *Partially treated sewage flows from Winnipeg plant*
CBC News
Posted: Nov 2, 2011 4:36 PM CT
http://www.cbc.ca/news/canada/manitoba/story/2011/11/02/winnipeg-wastewater-
plant-problem.html

Appendix i

The Winnipeg Sun ■ FRIDAY, DECEMBER 23, 2011

Sewage gets full treatment

Winnipeg is no longer spewing raw sewage into the Red River.

City officials said the effluent leaving Winnipeg's South End Water Pollution Control Plant is now fully treated.

From Oct. 7 to early November, it was only being treated half as well as it normally would, after the bacteria used in the city's biological treatment process died off en masse.

Beneficial bacteria

After Nov. 17, not long after city staff brought the issue to the mayor's attention, daily shipments of up to 10 loads of beneficial bacteria from Winnipeg's North End treatment plant as well as chlorine and polymer treatments were used to bring the water to nearly normal standards.

But before the issue was fixed, more than one billion litres of the partially treated sewage made it into the river.

The cause of the massive die-off of the beneficial bacteria is still unknown. Last month, Mayor Sam Katz said it may have been linked to the illegal dumping of diesel fuel nearby.

A city spokesman told media new checks are in place to prevent a reoccurrence.

— Winnipeg Sun

Appendix ii

Partially treated sewage flows from Winnipeg plant
CBC News
Posted: Nov 2, 2011
http://www.cbc.ca/news/canada/manitoba/story/2011/11/02/winnipeg-wastewater-plant-problem.html

More than one billion litres of partially treated sewage have flowed into the Red River over the past few weeks, due to a major glitch at a Winnipeg waste treatment plant, according to city officials.

One of the four sewage treatment processes at the South End Water Pollution Control Plant has not been working since Oct. 7, and city staff are stumped as to how and why it broke down.

As a result of the malfunction, effluent coming from the plant — which flows into the Red River — is currently being treated to just 50 per cent of how it would normally be treated, officials said Wednesday.

"Right now we are discharging wastewater to the Red River from the effluent of the plant that is in excess of our license requirement," Mike Shkolny, the city's manager of engineering services, told reporters.

Index   **Page**

Chapter 1 - Drinking Water Safety

The authors started by looking at the EPA standards for drinking water.

After all, as cited, "water systems must use EPA-approved analytical methods when analyzing samples to meet federal monitoring requirements or to demonstrate compliance with drinking water regulations."[1]

As further note, "all drinking water supplies should be analysed routinely for coliform bacteria and the general bacterial population."[2]

"A tolerable daily intake (TDI) is the amount of a substance that can be consumed from all sources each day by an adult, even for a lifetime, without any significant increased risk to health, based on current knowledge."[3]

And the "maximum acceptable concentration (MAC) of Escherichia coli in public, semi-public, and private drinking water systems is none detectable per 100 mL."[4]

It's a pro-active requirement of governments to ensure drinking water is safe.

Five case studies emphasize this importance, and the multitude of competing factors that might "pollute" drinking water[5]:

Case study – County of Oxford, Ontario[6]

As noted, "a watershed is simply the land that water flows across or through on its way to a common stream, river, or lake. A watershed can be very large (e.g. draining thousands of square miles to a major river or lake or the ocean), or very small, such as a 20-acre watershed that drains to a pond. A small watershed that nests inside of a larger watershed is sometimes referred to as a sub watershed."[7]

As further reported, "the contaminated water tragedy that struck Walkerton, Ontario, Canada in 2000 underlines the current view that watershed management is less about managing natural resources and more about managing human activities that affect those resources. We have a terrible tragedy here."[8]

"With those words, then Ontario Premier Mike Harris waded into the Walkerton, Ont., water crisis on Friday, May 26, 2000. He addressed a crowd of reporters and residents in the normally quiet town in Ontario's rural heartland – a part of the province that normally gears up for a flood of fun seekers at this time of year."[9]

Instead, Walkerton began the transition into the town "where those kids died from E. coli."[10]

"It's not what anyone wanted, but it was the end result."[11]

"Reporters from around North America descended on the area, trying to get to the bottom of Canada's worst-ever outbreak of E. coli contamination."[12]

"Seven people died from drinking contaminated water. Hundreds suffered from the symptoms of the disease, not knowing if they too would die."[13]

"According to the local medical officer of health, it all could have been prevented."[14]

"Dr. Murray McQuigge stunned the country with his revelation on CBC Radio on May 25, 2000 that the Walkerton Public Utilities Commission knew there was a problem with the water several days before they told the public."[15]

"Walkerton Inquiry Commissioner Dennis O'Connor recognized that source water protection is primarily an exercise in land use planning."[16]

In looking at Oxford County, which is located in the heart of southwestern Ontario, at the crossroads of Highways 401 and 403, one can see the competing sources for and potential contaminates of ground water.

"Characterized by rolling hills and productive farmland, the area is designated rural and agriculture is the primary activity."[17]

"Aggregate extraction is another major activity, with 40 percent of the province's limestone quarried here. Other area activities with the potential to influence source water quality and quantity are manufacturing (mainly automotive), construction, small businesses, and highway maintenance."[18]

"Virtually 100 percent of the county's water supply comes from groundwater sources."[19]

"About 100 000 people obtain their drinking water in this area: 70 000 residents are on municipal water systems, and 30 000 have private wells."[20]

"In keeping with the activities that take place in the county, agrochemicals and manure (especially from large livestock operations) are potential contaminants of source water. Industrial leaks and spills, road salt, road spills, and previous and existing industrial land uses also pose a threat to source water quality."[21]

"During the early 1990s, the County of Oxford undertook a major review of its one-tier Official Plan, a land use policy document laying out land use strategies for housing, industry, and commercial development, as well as providing for the protection of agricultural, aggregate, and environmental resources. As a result of this review, new policies for groundwater protection were developed and implemented through the environmental protection sections of the Plan."[22]

The County incorporated all five aquifer areas into the Official Plan, under the designation "Groundwater Recharge Areas." These areas are protected by restricting incompatible land uses. Some land uses fall into the category of "conditional uses," which include municipal and industrial landfills; lagoons or other putrescible waste disposal facilities; asphalt and concrete batching plants; industrial and commercial uses, other than farms, involving the storage or processing of chemical and/or petroleum products; gasoline or oil depots; service stations, and vehicle salvage, maintenance, and service yards. Uses of this type already in existence are allowed to continue through recognition in the zoning by-law of the area municipality.[23]

"To assess general groundwater quality, 84 shallow overburden wells and 83 bedrock wells were sampled between April and August 2000. Samples were analyzed for general chemical and bacteriological parameters; samples from selected wells were also analyzed for metals, volatile organic compounds, and pesticides."[24]

"Bacteriological results showed a higher incidence of total coliform and E. coli in the shallow wells."[25]

"Shallow aquifer groundwater was slightly basic on average, with moderate levels of nitrate, chloride, total dissolved solids, and conductivity. It generally contained low levels of dissolved organic carbon, sulphate, and metals. The higher levels of chloride and nitrate in shallow aquifer groundwater reflect a generally higher susceptibility of the overburden aquifer to surficial contamination by fertilizers, septic waste, and road salt."[26]

"In comparison, bedrock aquifer groundwater was slightly higher levels of conductivity, total dissolved solids, and sulphate were most likely a natural characteristic of the bedrock aquifer."[27]

To help protect "sensitive areas around a municipal water supply well and vulnerable aquifer areas are protected from potential water contamination", land use is restricted according to the North American Industry Classification System.[28]

The North American Industry Classification System (NAICS), "is the standard used by Federal statistical agencies in classifying business establishments for the purpose of collecting, analyzing, and publishing statistical data related to the U.S. business economy."[29]

NAICS "was developed under the direction and guidance of the Office of Management and Budget (OMB) as the standard for use by Federal statistical agencies in classifying business establishments for the collection, tabulation, presentation, and analysis of statistical data describing the U.S. economy. Use of the standard provides uniformity and comparability in the presentation of these statistical data. NAICS is based on a production-oriented concept, meaning that it groups establishments into industries according to similarity in the processes used to produce goods or services."[30]

Part of the protection of ground water was exemplified by this incident:

"In 1997 an application for a building permit for two hog barns with a total barn capacity of 10 000 feeder hogs was received by the Township of South-West Oxford."[31]

"The proposed location of the barns was relatively close to the main Woodstock well field and to the Village of Sweaburg supply wells."[32]

"Later mapping of the capture zones indicated that the proposed barns would have fallen within these zones."[33]

"Area residents and Woodstock residents and public utility representatives resisted the proposal, citing concerns about the effect of such a large operation on the drinking water supply, as well as about odour."[34]

"In response to these concerns, the County passed an Interim Control By-law that prohibited the establishment of new livestock operations over 500 livestock units in size (2000 feeder hogs, for example)."[35]

Case study - North Saskatchewan Watershed in Alberta[36]

As noted, "no matter where we live or work, we are in a watershed teeming with unique, inter-related natural processes. In mountain upland areas, there are unique blends of climate, geology, hydrology, soils, and vegetation shaping the landscape, with waterways often cutting down steep slopes. In an upland plains area, you find grassy plains, hardy vegetation, and slower moving, meandering streams and rivers. In the coastal area, where oceans meet land, there are again different blends of features and processes shaping the environment. In lowland areas between upland and coastal waters, where tidal wetlands are prevalent, processes serve entirely different functions."[37]

As cited, "with a land area of 80 000 square kilometres, making up about 12.5 percent of Alberta's total area, the Alberta portion of the North Saskatchewan Watershed provides an example of a very large watershed. The Alberta portion of the North Saskatchewan Watershed is located in central Alberta and comprises the main stem of the North Saskatchewan River and its major tributaries, including the Battle River Basin and Sounding Creek."[38]

"Most of the watershed's population resides in the greater Edmonton area. The Alberta portion of the North Saskatchewan Watershed is also home to 10 provincial parks, three national parks, three ecological reserves, eight provincial grazing reserves, two wilderness areas, 18 Aboriginal reserves representing 14 tribes, and three Métis settlements."[39]

"Much of the urban population in the Alberta portion of the North Saskatchewan Watershed receives treated drinking water from the North Saskatchewan River and its tributaries (surface water)."[40]

"There are many types of water demands on the North Saskatchewan River in Alberta. These include hydroelectric generation, human consumption, oil and gas extraction, mining, and agricultural uses such as irrigation and livestock watering."[41]

"A large industrial base in the Edmonton area withdraws water from the North Saskatchewan River for cooling and process waters. Oil and gas activities, including deep-well injection for enhanced oil recovery."[42]

Case study – Amherst, Nova Scotia[43]

"Nova Scotia's Environment Act is the main tool for managing water resources in the province. Municipalities can request that the Department designate their drinking water source areas (both surface and groundwater) as protected water areas under the Water Act. Currently 24 of the province's 82 municipal water supplies are designated as protected water areas."[44]

"The County of Cumberland is one of the largest rural municipalities in Nova Scotia. Bounded by the Bay of Fundy and the Northumberland Strait, it covers the Isthmus of Chignecto which joins Nova Scotia to New Brunswick. The economic base of Cumberland County is built on forestry, agriculture, and fishery. There are numerous small communities throughout the County, but the most densely developed and rapidly growing residential areas are in the Town of Amherst, which provides an employment and service centre for the County."[45]

"The protected water area of the North Tyndal Wellfield, the source of drinking water for both the Town and parts of the County, is located about 15 kilometres east of Amherst, between Amherst and Tidnish."[46]

"The North Tyndal Wellfield consists of four production wells with a combined average pumping rate of 8 400 litres per minute. The wells extract water from a bedrock aquifer consisting of fractured sedimentary rocks of the Pictou Group."[47]

"In keeping with the activities that take place in the wellfield, there is the potential for the following contaminants to affect groundwater quality:

•*pathogens, including bacteria and viruses*

•*nutrients, including nitrates and phosphates*

•*salt, including sodium chloride and calcium chloride*

•*hydrocarbons, such as gasoline, fuel diesel, and lubricating oil, as well as individual components of these products, including benzene, toluene, ethylbenzene, and xylene*

•*biocides, including pesticides and herbicides, as well as possible breakdown products*

•other organic compounds, such as those used in solvents, paints, fertilizers, and wood preservatives, as well as possible breakdown products."[48]

"Groundwater flow within the wellfield site was modeled using the U.S. Geological Survey models MODFLOW and MODPATH and the U.S. Environmental Protection Agency's Well Head Protection Area."[49]

Case study – Edmundston, New Brunswick[50]

As noted, "working with your watershed also means understanding how most human activities in the watershed can occur in harmony with natural processes. Communities located along streams and rivers, for example, are faced with very basic choices: they can learn how the river functions and learn to draw benefits from it while staying out of harm's way—or, they can try to significantly change the river's behavior in order to accomplish their plans. It may be feasible to change the way a river acts, but this usually means taking on costly and never-ending maintenance of those man-made changes; and, despite all the maintenance, communities may remain still vulnerable to floods and other disasters. In contrast, a community that has made sensible decisions on activities near the river can avoid a costly maintenance burden while sustaining their community's use and enjoyment of a healthy river system."[51]

"New Brunswick's Watershed Protection Program was initiated in the late 1980s as means of protecting surface water sources of drinking water. In 1990, the Watercourse Setback Designation Order was put into place, giving protected status to 31 surface watersheds that supplied potable water to 24 municipalities. The program was mandatory, and all 31 watersheds were designated as protected areas. About 60 percent of New Brunswick's population (about 430 000 people) get their drinking water from groundwater sources. The Municipality of Edmundston is located in northwestern New Brunswick, near the New Brunswick/Maine border and about 15 kilometres from the Quebec border. The outlying areas are predominantly farming communities with smaller populations (500 to 5000 people)."[52]

"The surficial geology of the area generally comprises Wisconsonian and Late Wisconsonian aged sediments. Striations indicate that the last ice sheet retreated from the west-northwest, leaving glacio-fluvial deposits of mainly sands and gravels deposited by meltwater streams. The valleys of Three Mile Brook, the Madawaska River, the Iroquois River, and some smaller watercourses have been infilled with these glacio-fluvial materials. These surficial materials overlie a bedrock comprising igneous, metamorphic, and sedimentary rocks of Lower Devonian age."[53]

"The City of Edmundston Municipal Watershed area is composed of two separate watersheds: the Iroquois River Watershed, with an area of 19 684 hectares, and the Blanchette Brook Watershed, with an area of 3636 hectares. The watershed areas extend north of the city boundary for a distance of about 30 kilometres, across the border with New Brunswick, and into Quebec. A small portion (about 90 hectares) of the Iroquois River Watershed is located within city boundaries and is thus under the jurisdiction of the City."[54]

"The City currently draws its supply of potable water from two wellfields within these watersheds: the Iroquois Wellfield, located 6 kilometres to the north of the city centre, and another at Verret, about 3 kilometres west of the City."[55]

"In the late 1980s, incidents of contamination of the water supply in various parts of the province raised concerns over the protection of public water supplies. In the late 1980s, watershed designation studies were carried out for each of the watersheds associated with sourcing the Municipality of Edmundston's public drinking water supply."[56]

"The raw water quality of the Iroquois River and Blanchette Brook was found to be within the Canadian Drinking Water Guidelines with respect to chemical constituents. However, it consistently exceeded the standards for bacteriological counts. On one occasion, the fecal coliform density was found to be in excess of 4000 per millilitre, and the total coliforms were in excess of 11 000 per 100 millilitres."[57]

Case study – Powell River, British Columbia[58]

As noted, "one should never look for a rigid, step-by-step 'cookbook recipe' for watershed management. One size does not fit all; different regions of the country have watersheds that function in very different ways. Even neighboring watersheds can have major differences in geology, land use, or vegetation that imply the need for very different management strategies. Different communities vary in the benefits they want from their watersheds. Moreover, watersheds change through time. Eastern watersheds cleared of their forests in the first half of the 20th century had specific management needs during regrowth in the second half of the century, but management needs will likely change again in the 21st century. Changes can even occur on more immediate time scales, due for example to the appearance of a serious forest pest or disease, a change in water use patterns, or the arrival of a new community industry or enterprise. Watershed management is a dynamic and continually readjusting process that is built to accommodate these kinds of changes."[59]

In June 2002, the provincial government announced a new plan that would enhance drinking water protection through "source protection, monitoring, assessments, and infrastructure investment. Powell River is located about 135 kilometres north of Vancouver on British Columbia's Sunshine Coast. Land within the associated Haslam Lake-Lang Creek watershed is mainly Crown Provincial Forest, with some small areas of private land along the lower reaches of Lang Creek, as well as occasional parcels in the lower Haslam Lake watershed areas."[60]

"The entire water supply in this watershed is derived from surface water. The Corporation of the District of Powell River has its main water supply intake on the southwestern tip of Haslam Lake. This intake supplies 100 percent of the water used by the Townsite, Cranberry, and Westview areas, which have a population of about 12 500. The district has two licenses on the lake, one for 4 913 630 000 gallons per year and the other for 14 400 acre-feet of water storage."[61]

"Logging has been an important activity in the watershed since the late 1800s to early 1900s."[62]

"By the early 1990s, the ability of the Haslam Lake and Lang Creek systems to supply an adequate quantity of water for domestic and fisheries uses had become a concern. There were also concerns over potential deterioration of water quality of community water supplies due to ongoing and proposed industrial, resource-extraction, and recreational activities in the watershed."[63]

"Activities on these lands that posed a potential risk to water quality included: land clearing along creek or lake foreshores; livestock utilization of foreshore areas and disposal of livestock manure; use and disposal of fertilizers, pesticides, and herbicides associated with agricultural production and property maintenance; water-based recreation; and disposal of industrial and household waste."[64]

The impact of safe drinking water was emphasized by news articles like this, "1,775 boil-water advisories in Canada require action"[65]:

"More than 1,700 boil-water advisories are in effect in communities across the country, according to a new investigative report by the Canadian Medical Association."[66]

"A boil-water advisory means that water is contaminated and unfit to drink without boiling. Often, the report says, a community's chlorination or disinfection systems fail to work, leading to advisories."[67]

"Although boil-water advisories are often associated with native communities, 93 First Nations had advisories in place as of Feb. 29, 2008, while 1,766 advisories outside of these communities were in place in Canada at the end of the following month, the report said."[68]

"Advisories are intended to be a precautionary measure in the public health tool kit, but given that some have been in place for at least five years, they are apparently being used as a Band-Aid substitute for treatment," write the authors in the report."[69]

"Ninety people in Canada die and another 90,000 get sick from drinking contaminated water each year, the report said."[70]

A Google map showing the "boil water" advisories for Canada, in 2011 appears as **Appendix 1** [cited below].[71]

Footnotes

1. *Drinking Water Analytical Methods*
http://water.epa.gov/scitech/drinkingwater/labcert/analyticalmethods.cfm

2 - 3. *Guidelines for Canadian Drinking water*
http://docs.google.com/viewer?a=v&q=cache:OwJdLz9-vywJ:xnet.rrc.mb.ca/rcharney/water%2520quality.pdf+canadian+drinking+water+guidelines&hl=en&gl=ca&pid=bl&srcid=ADGEEShmUBJ6BCXV_LDJUfOBPwlVhfkwPrR5rK72LhfJMopt4-

PfTs0h4j_mJNP41zRI6eTMdjxVL4-
VSHANgJGbLFup0z_JYa7h6rUBfE556UdOlrxBk6g6hDaHTywCein
y3wyPqrF8&sig=AHIEtbSS8c-jHJnO_6f3hJFCGbUW1MLiMA

4. *Bacteriological guidelines*
http://www.hc-sc.gc.ca/ewh-semt/pubs/water-eau/2010-sum_guide-
res_recom/index-eng.php

5. *Drinking Water*
http://www.ec.gc.ca/eau-water/default.asp?lang=En&n=AAD01CB4-1

6. *Case study – County of Oxford, Ontario*
http://www.ec.gc.ca/eau-water/default.asp?lang=En&n=F33CB10C-1

7. *Watershed Academy Web: Online Training in Watershed
Management*
http://cfpub.epa.gov/watertrain/index.cfm

8 - 15. *Canada's worst-ever E. coli contamination*
CBC News Online | Updated Dec. 20, 2004
http://www.cbc.ca/news/background/walkerton/

16 - 28. *Case study – County of Oxford, Ontario*
http://www.ec.gc.ca/eau-water/default.asp?lang=En&n=F33CB10C-1

29 - 30. *The North American Industry Classification System*
http://www.census.gov/eos/www/naics/

31 - 35. *Case study – County of Oxford, Ontario*
http://www.ec.gc.ca/eau-water/default.asp?lang=En&n=F33CB10C-1

36. *Case study - North Saskatchewan Watershed in Alberta*
http://www.ec.gc.ca/eau-water/default.asp?lang=En&n=CC09EDA0-1

37. Watershed Academy Web: Online Training in Watershed
Management
http://cfpub.epa.gov/watertrain/index.cfm

38 - 42. *Case study - North Saskatchewan Watershed in Alberta*
http://www.ec.gc.ca/eau-water/default.asp?lang=En&n=CC09EDA0-1

43 - 48. *Case study – Amherst, Nova Scotia*
http://www.ec.gc.ca/eau-water/default.asp?lang=En&n=87D5AFD0-1

48. *Case study – Amherst, Nova Scotia*
http://www.ec.gc.ca/eau-water/default.asp?lang=En&n=87D5AFD0-1

Also see: MODFLOW
http://en.wikipedia.org/wiki/MODFLOW

MODFLOW is the U.S. Geological Survey modular finite-difference flow model, which is a computer code that solves the groundwater flow equation. The program is used by hydrogeologists to simulate the flow of groundwater through aquifers. The code is free software, written primarily in Fortran, and can compile and run on DOS, Windows or Unix-like operating systems.

Also see: MODPATH
http://www.xmswiki.com/xms/GMS:MODPATH

MODPATH is a particle tracking code that is used in conjunction with MODFLOW. After running a MODFLOW simulation, the user can designate the location of a set of particles. The particles are then tracked through time assuming they are transported by advection using the flow field computed by MODFLOW. Particles can be tracked either forward in time or backward in time. Particle tracking analyses are particularly useful for delineating capture zones or areas of influence for wells.

50. *Case study – Edmundston, New Brunswick*
http://www.ec.gc.ca/eau-water/default.asp?lang=En&n=DFF500EF-1

51. *Watershed Academy Web: Online Training in Watershed Management*
http://cfpub.epa.gov/watertrain/index.cfm

52 - 57. *Case study – Edmundston, New Brunswick*
http://www.ec.gc.ca/eau-water/default.asp?lang=En&n=DFF500EF-1

58. *Case study – Powell River, British Columbia*
http://www.ec.gc.ca/eau-water/default.asp?lang=En&n=9B723C7A-1

59. *Watershed Academy Web: Online Training in Watershed Management*
http://cfpub.epa.gov/watertrain/index.cfm

60 - 64. *Case study – Powell River, British Columbia*
http://www.ec.gc.ca/eau-water/default.asp?lang=En&n=9B723C7A-1

65 - 70. *1,775 boil-water advisories in Canada require action: report*
CBC News, Last Updated: Monday, April 7, 2008
http://www.cbc.ca/news/health/story/2008/04/07/boil-advisory.html

71. *Google Map of Canada displaying numbers of water advisories in each province and territory.* ... *November 12, 2011*
http://www.water.ca/map-graphic.asp

Appendix 1

Water advisories

Google Map of Canada displaying numbers of water advisories in each
province and territory. ... November 12, 2011
http://www.water.ca/map-graphic.asp

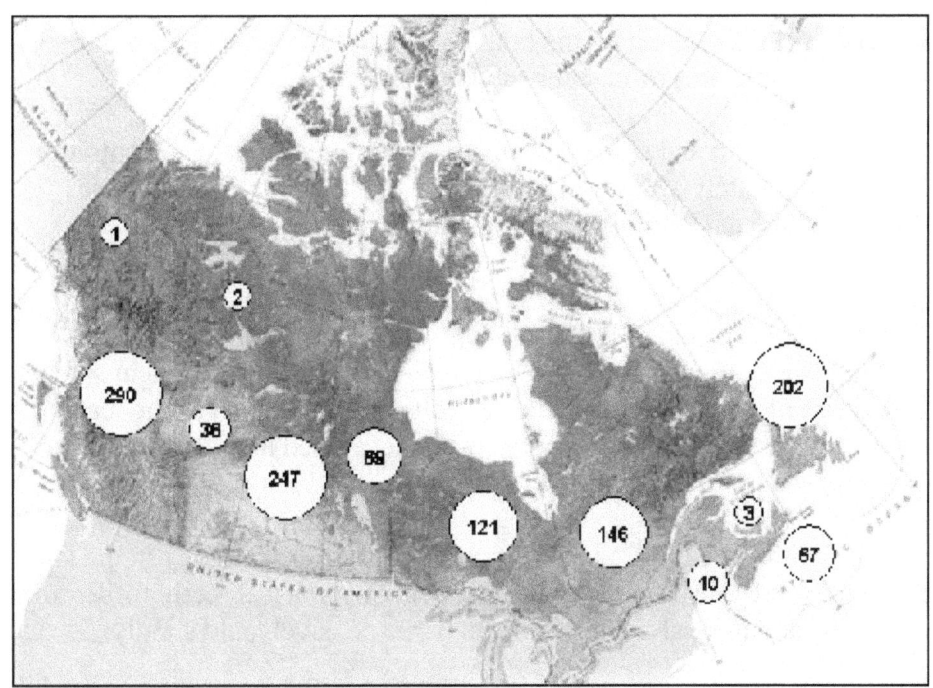

Chapter 2 - The Dangers of E. Coli

As cited, "Escherichia coli is a member of the coliform group, part of the family Enterobacteriaceae, and is described as a facultative anaerobic, Gram-negative, non-spore-forming, rod-shaped bacterium that possesses the enzyme β-glucuronidase."[1]

"As a member of the Enterobacteriaceae family, E. coli is naturally found in the intestines of humans and warm-blooded animals."[2]

"Unlike other bacteria in this family, E. coli does not usually occur naturally on plants or in soil and water."[3]

"Within human and animal faeces, E. coli is present at a concentration of approximately 109 per gram (Edberg et al., 2000) and comprises about 1% of the total biomass in the large intestine (Leclerc et al., 2001)."[4]

In fact, Edberg et al. (2000) mentions that Escherichia coli "in the 1890s was chosen as the biological indicator of water treatment safety. Because of method deficiencies, E. cult surrogates such as the fecal coliform' and total coliforms tests were developed and became part of drinking water regulations."[5]

And ,"with the advent of the Defined Substrate Technology in the late 1980s, it became possible to analyse drinking water directly for E. coli (and, simultaneously, total coliforms) inexpensively and simply."[6]

Finally, "E. coli survives in drinking water for between 4 and 12 weeks, depending on environmental conditions (temperature, microflora, etc.)."[7]

Hence, it is important to disinfect drinking water, to guard against such contaminants.

As cited, "the purpose of treating drinking water is to provide a product that is microbiologically and chemically safe for consumption."[8]

"In all public and semi-public systems applying disinfection, a disinfectant residual should be maintained throughout the distribution system at all times."[9]

"Maintenance and monitoring of a residual disinfectant offer two benefits. First, a disinfectant residual will limit the growth of organisms within the system and may afford some protection against contamination from without; second, the disappearance of the residual provides an immediate indication of the entry of oxidizable matter into the system or of a malfunction of the treatment process."[10]

"Most drinking water treatment plants in Canada use chlorine as a disinfectant. The use of chlorine in the treatment of drinking water has virtually eliminated waterborne diseases, because chlorine can kill or inactivate most microorganisms commonly found in water [see **Appendix 2a, 2b & 2c**, cited below]."[11]

Furthermore, "the majority of drinking water treatment plants in Canada use some form of chlorine to disinfect drinking water: to treat the water directly in the treatment plant and/or to maintain a chlorine residual in the distribution system to prevent bacterial regrowth."[12]

"In addition, sample locations should be chosen to represent all areas of the distribution system, taking into account geographical location, age and materials of the water main, structural integrity of the distribution system, water storage, and retention times."[13]

In fact, "the maximum acceptable concentration (MAC) of Escherichia coli in public, semi-public, and private drinking water systems is none detectable per 100 mL [see **Appendix 3a & 3b**, cited below]."[14]

"Testing for E. coli should be carried out in all drinking water systems. The number, frequency, and location of samples for E. coli testing will vary according to the type and size of the system and jurisdictional requirements."[15]

In the size of Winnipeg, Canada it is a relatively small city, but one would surmise that adequate precautions and sampling is carried out in their drinking water distribution system?

In terms of detecting E. coli organisms, as cited, "currently, three methods are routinely used to detect E. coli organisms in water: presence-absence (P-A), which is a qualitative test, and membrane filter (MF) and multiple tube fermentation (MTF), which are both quantitative tests.[16]

"Methods that detect and confirm the presence of E. coli in a single step use the enzyme β-glucuronidase, a unique constitutive enzyme found in E. coli, Shigella spp., and some Salmonella spp., but rarely present in other coliforms (Manafi et al., 1991)."[17]
As further cited by Geissler et. a. (2000), "new enzymatic methods have been developed. The use of media containing chromogenic or

fluorogenic substrates for the enzymes β-galactosidase (Lac) and β-glucuronidase (Gus) for simultaneous detection of coliforms and E. coli is increasing."[18]

As further cited, the P-A [Presence - Absence] "test was developed as a more sensitive, economical, and efficient means of analysing drinking water samples (Clark and Vlassoff, 1973)."[19]

"This procedure is currently the preferred method, in many jurisdictions, for verifying the bacteriological safety of public drinking water supplies (i.e., the absence of E. coli)."[20]

"Essentially, the P-A test is a modification of the MTF procedure in which only one analysis bottle per sample is used. This method can be used with either enzyme-based media, such as media based on defined substrate technology, or presumptive coliform media (e.g., using lauryl tryptose broth), with follow-up E. coli confirmation."[21]

"Commercial test kits using defined substrate technology have been developed. Studies performed on the effectiveness of the commercial tests compared with classical MTF and MF approaches showed that the commercial kits were usually as sensitive as the MTF approach for the detection of E. coli, and sometimes more sensitive for the detection of total coliforms (Rompré et al., 2002)."[22]

Rompré et al. (2002) reported "approved traditional methods for coliform detection include the multiple-tube fermentation (MTF) technique and the membrane filter (MF) technique using different specific media and incubation conditions."[23]

"These methods have limitations, however, such as duration of incubation, antagonistic organism interference, lack of specificity and poor detection of slow-growing or viable but non-culturable (VBNC) microorganisms."[24]

"Nowadays, the simple and inexpensive membrane filter technique is the most widely used method for routine enumeration of coliforms in drinking water."[25]

"The enzymes β-d galactosidase and β-d glucuronidase are widely used for the detection and enumeration of total coliforms and Escherichia coli, respectively. Many chromogenic and fluorogenic substrates exist for the specific detection of these enzymatic activities, and various commercial tests based on these substrates are available. Numerous comparisons have shown these tests may be a suitable alternative to the classical techniques."[26]

In fact, Ramteke et. al. (1993) analyzed "one thousand three-hundred and ninety-four drinking water sources comprising ground water, surface water and piped supplies were tested in order to compare the presence-absence (P-A) test with standard MPN method to detect coliforms as indicators of water quality."[27]

The results, "out of 1394 samples, 1074 (77.04%) and 1030 (74.88%) were positive by the MPN and P-A test, respectively. The P-A test detected 96% of the positives detected by the MPN test."[28]

In addition, "in comparative tests using lactose-based media, the P-A method was shown to be at least as sensitive as the MF and MTF techniques for the recovery of both total coliforms and E. coli."[29]

"Technically, the P-A test is simpler than the MF and MTF procedures. Initial per-sample analysis time is less than 1 minute, and, since the allowable level of E. coli in drinking water is none per 100 mL, qualitative results are sufficient for protecting public health." [30]

Footnotes

1 - 3. *Significance of E. coli in drinking water*
http://www.hc-sc.gc.ca/ewh-semt/pubs/water-eau/escherichia_coli/significance-importance-eng.php

3. *Significance of E. coli in drinking water*
http://www.hc-sc.gc.ca/ewh-semt/pubs/water-eau/escherichia_coli/significance-importance-eng.php

Also see: Edberg, S.C., Rice, E.W., Karlin, R.J., and Allen, M.J. (2000) *Escherichia coli: the best biological drinking water indicator for public health protection.* J. Appl. Microbiol., 88: 106S-116S.

Also see: Leclerc, H., Mossel, D.A.A., Edberg, S.C., and Struijk, C.B. (2001) *Advances in the bacteriology of the coliform group: their suitability as markers of microbial water safety.* Annu. Rev. Microbiol., 55: 201-234.

5 - 7. Edberg, S.C., Rice, E.W., Karlin, R.J., and Allen, M.J. (2000) *Escherichia coli: the best biological drinking water indicator for public health protection.* J. Appl. Microbiol., 88: 106S-116S.

8 - 13. *Role of disinfectant residuals in maintaining drinking water quality*

http://www.hc-sc.gc.ca/ewh-semt/pubs/water-eau/escherichia_coli/disinfectant-desinfectant-eng.php

14 - 15. *Escherichia coli: Bacteriological guidelines*
http://www.hc-sc.gc.ca/ewh-semt/pubs/water-eau/2010-sum_guide-res_recom/index-eng.php

16. *Role of disinfectant residuals in maintaining drinking water quality*
http://www.hc-sc.gc.ca/ewh-semt/pubs/water-eau/escherichia_coli/disinfectant-desinfectant-eng.php

17. *Role of disinfectant residuals in maintaining drinking water quality*
http://www.hc-sc.gc.ca/ewh-semt/pubs/water-eau/escherichia_coli/disinfectant-desinfectant-eng.php

See also: Manafi M, Kneifel W, Bascomb S., *Fluorogenic and chromogenic substrates used in bacterial diagnostics*. Microbiol Rev. 1991 Sep;55(3):335-48.

18. K. Geissler, M. Manafi, I. Amorós, J. L. Alonso, *Quantitative determination of total coliforms and Escherichia coli in marine waters with chromogenic and fluorogenic media*. Journal of Applied Microbiology, Volume 88, Issue 2, pages 280–285, February 2000

19. *Role of disinfectant residuals in maintaining drinking water quality*
http://www.hc-sc.gc.ca/ewh-semt/pubs/water-eau/escherichia_coli/disinfectant-desinfectant-eng.php

Also see: Clark JA & Vlassoff LT, *Relationships among pollution indicator bacteria isolated from raw water and distribution systems by the presence-absence (P-A) test*. Health Laboratory Science [1973, 10(3):163-72]

20 - 22. *Role of disinfectant residuals in maintaining drinking water quality*
http://www.hc-sc.gc.ca/ewh-semt/pubs/water-eau/escherichia_coli/disinfectant-desinfectant-eng.php

23 - 26. Annie Rompré, Pierre Servais , Julia Baudart, Marie-Renée de-Roubin, Patrick Laurent, *Detection and enumeration of coliforms in drinking water: current methods and emerging approaches*. Journal of Microbiological Methods, Volume 49, Issue 1, March 2002, Pages 31-54

27 - 28. P. W. Ramteke, S. P. Pathak, J. W. Bhattacherjee, K. Gopal and N. Mathur, *Evaluation of the presence-absence (P-A) test: A*

simplified bacteriological test for detecting coliforms in rural drinking water of India. Environmental Monitoring and Assessment, Volume 33, Number 1, 53-59, DOI: 10.1007/BF00546661, 1993.

29. *Role of disinfectant residuals in maintaining drinking water quality* http://www.hc-sc.gc.ca/ewh-semt/pubs/water-eau/escherichia_coli/disinfectant-desinfectant-eng.php

Also see: J A Clark and A H el-Shaarawi, *Evaluation of commercial presence-absence test kits for detection of total coliforms, Escherichia coli, and other indicator bacteria.* Appl. Environ. Microbiol. February 1993 vol. 59 no. 2 380-388

30. *Role of disinfectant residuals in maintaining drinking water quality* http://www.hc-sc.gc.ca/ewh-semt/pubs/water-eau/escherichia_coli/disinfectant-desinfectant-eng.php

Appendix 2a

Guidelines for Drinking Water

Guidelines for Canadian Drinking Water Quality - Chlorine Guideline
Technical Document, June 2009
ISBN: 978-1-100-13416-1; Cat. No.: H128-1/09-588E
HC Pub.: 4188
http://webcache.googleusercontent.com/search?q=cache:pWdkUzUK4
HcJ:www.hc-sc.gc.ca/ewh-semt/pubs/water-eau/chlorine-chlore/index-
eng.php+canada+water+biochemistry+field+test+kits&cd=5&hl=en&c
t=clnk&gl=ca&source=www.google.ca

Chemistry in aqueous media

When added to water, chlorine gas (Cl_2) dissolves rapidly and establishes an equilibrium with hypochlorous acid (HOCl), according to chemical equation (1):

$$Cl_2 + H_2O \leftrightarrow H^+ + Cl^- + HOCl \qquad (1)$$

Addition of NaOCl and $Ca(OCl)2$ to water achieves the same essential oxidizing agent, HOCl, according to chemical reactions (2) and (3) below (IARC, 1991; Connell, 1996; White, 1999), with the only difference being side reactions and end products:

$$NaOCl + H_2O \leftrightarrow NaOH + HOCl \qquad (2)$$

$$Ca(OCl)_2 + 2H_2O \leftrightarrow Ca(OH)_2 + 2HOCl \qquad (3)$$

HOCl then dissociates to negative hypochlorite ion (OCl^-) according to chemical equation (4):

$$HOCl \leftrightarrow H^+ + OCl^- \qquad (4) \qquad pK_a = 7.5$$

All of these chemical reactions, and thus disinfection effectiveness, are highly dependent upon the pH and temperature of the aqueous medium, which determine the extent of conversion between the three free chlorine species: Cl_2, HOCl, and OCl^-. HOCl is considered to be more effective at microbial inactivation and dominates at lower pH levels. For example, at a pH of 6.5 and temperatures of 0°C and 20°C, an aqueous solution of chlorine would contain about 95.5% and 92.4% HOCl, respectively (4.5% and 7.6% of OCl^-); at a higher pH of 8.5, the equilibrium shifts to 17.5% and 10.8% of HOCl (82.5% and 91.2% of OCl^-). The typical pH range of drinking water is between 6.5 and 8.5, and chlorination of drinking water at pH levels below 8 provides maximum disinfection efficiency (White, 1999; IPCS, 2000; WHO, 2004). However, efficiency can also be increased by increasing contact time, concentration, or temperature (U.S. EPA, 1999b, 2007).

Appendix 2b

Chloride Dioxide

CHLORINE DIOXIDE:
http://home.windstream.net/mikeric/Odor/clo2.htm

Cited *ver batim*: *"What You Can't Tell From the Name*

While chlorine dioxide has chlorine in its name, its chemistry is radically different from that of chlorine.

The way it works is almost magical. It has to do with the way electrons interact with one another. As we all learned in high school chemistry, we can mix two compounds and create a third that bears little resemblance to its parents.

For instance: Mix two parts of hydrogen gas with one of oxygen and liquid water is the result.

Mix equal parts of caustic soda (commonly called lye, a part of everyday soap) and hydrochloric acid (which will dissolve iron) and you get table salt and water. And for chlorine dioxide, mix one part chlorine gas with two parts of oxygen.

We should not be misled by the fact that chlorine and chlorine dioxide share a word in common.

Hydrogen is in both water and in hydrogen cyanide. The latter can be a deadly poison.

At room temperature, chlorine is a greenish-yellow poisonous gas. When added to water, however, chlorine combines with water to form hypochlorous acid that then ionizes to form hypochlorite ion - 'bleach'

Regarding bleaching, chlorine dioxide and chlorine -- because of their fundamentally different chemistries -- react in distinct ways with organic compounds, and as a result generate very different byproducts.

It is this difference that explains the superior environmental performance of chlorine dioxide in paper making and scrubbers.

Technically speaking, both chlorine and chlorine dioxide are oxidizing agents -- electron receivers.

Chlorine has the capacity to take in two electrons, whereas chlorine dioxide can absorb five.

This property, along with the complex, but well known, ways chlorine combines with lignin (the cellular adhesive in wood tissue), explains the basic difference between the two compounds.

In the chlorine-based bleaching process, about 10 percent of the chlorine combines directly with lignin which has "aromatic" components.

Aromatic compounds have atoms arranged in rings, and they may have other atoms, such as chlorine, attached to these rings.

Within the group of chlorinated aromatics, which can be toxic to some organisms, are the infamous dioxins.

Chlorine dioxide's behavior as a bleaching agent is quite dissimilar. Instead of combining with the aromatic rings, chlorine dioxide breaks these rings apart.

In addition, as the use of chlorine dioxide increases, the generation of chlorinated organics falls dramatically.

Chlorine dioxide's chemistry explains why it is such an effective oxidant, or bleaching agent.

It is 2.5 times more active than chlorine gas, and much more selective.

Chlorine dioxide attacks the lignin, but does not react with the desired cellulose in wood tissue. It is cellulose -- the tree's fiber -- that provides the strength in the final paper products.

These advantages make chlorine dioxide the preferred environmental standard for eliminating toxic substances in mill waste water and scrubbers.

Chlorine dioxide is a neutral compound of chlorine in the +IV oxidation state.

It has a boiling point of 11 degrees C at atmospheric pressure.

The liquid is denser than water, and the gas is denser than air.

The molecule is polar with the oxygen atoms separated by 116.5 degrees.

Water (H_2O) is also polar (105 degrees).

The presence of organic matter, nutrients, and microorganisms in the output of sewage treatment plants is measured by three tests: coliform count, algal count, and biochemical oxygen demand (BOD).

The coliform count describes the number of E. coli (the characteristic bacteria in animal wastes) present.

The algal count is a biological test for microorganisms other than bacteria and viruses which may be present.

The BOD measures the volume of oxygen gas taken up by a given amount of water in five days at 20 degrees C, (remember, there is an ultimate test of BOD).

The biochemical oxygen demand analysis is an attempt to simulate the effect a waste will have on the dissolved oxygen of a stream by a laboratory test.

The BOD test gives an indication of the amount of oxygen needed to stabilize or biologically oxidize the waste.

The advantage of the BOD test is that it measures only the organics which are oxidized by the bacteria.

The disadvantage is the 5 day time lag and the difficulty in obtaining consistent repetitive values.

The results of the COD (chemical oxygen demand) tests are usually higher that the corresponding BOD test for the following reasons:

Many organic compounds which are dichromate oxidizable are not biochemically oxidizable.

Certain inorganic substances, such as sulfides, sulfites, thiosulfates, nitrites and ferrous iron are oxidized by dichromate, creating an inorganic COD, which is misleading when estimating the organic content of the wastewater.

The BOD results may be affected by lack of seed acclimation, giving erroneously low readings. The COD results are independent of seed acclimation.

In general, chlorine dioxide has been found to produce fewer organic byproducts with naturally occurring dissolved organic material.

Chlorine dioxide is an explosive gas, but is stable in water in the absence of light and elevated temperatures ... which is just what we do.

ClO_2 is capable of oxidizing iron and manganese, removing color, and lowering THM (Trihalomethanes) formation potential.

It also oxidizes many organic and sulfurous compounds that cause off-tastes and odors.

Chlorine dioxide is a green-yellow gas that decomposes readily and with explosive force to chlorine and oxygen. It is, therefore, usually manufactured on-site.

Chlorine dioxide is a more powerful biocide than free chlorine but does not persist as long as chlorine."

Appendix 2c

Chlorine
http://www.gewater.com/handbook/cooling_water_systems/ch_27_chlorine.jsp

Although chlorine is beneficial for many uses, its use carries safety and environmental concerns.

Chlorine in its gaseous state was discovered by Karl W. Scheele in 1774 and identified as an element by Humphrey Davy in 1810. Chlorine gas is greenish-yellow, and its density is about 2 times that of air. When condensed, it becomes a clear, amber liquid with a density about 1 times greater than water. One volume of liquid chlorine yields approximately 500 volumes of chlorine gas, which is neither explosive nor flammable. Like oxygen, chlorine gas can support the combustion of some substances. Chlorine reacts with organic materials to form oxidized or chlorinated derivatives. Some of these reactions, such as those with hydrocarbons, alcohols, and ethers, can be explosive. The formation of other chlorinated organics, specifically trihalomethanes (THM), poses an environmental threat to public drinking water supplies.

Chlorine gas is also a toxic respiratory irritant. Airborne concentrations greater than 3-5 ppm by volume are detectable by smell, and exposure to 4 ppm for more than 1 hr can have serious respiratory effects. Because chlorine gas is denser than air, it stays close to the ground when released. The contents of a 1-ton cylinder of chlorine can cause coughing and respiratory discomfort in an area of 3 square miles. The same amount concentrated over an area of 1/10 square mile can be fatal after only a few breaths.

Chlorine is generated commercially by the electrolysis of a brine solution, typically sodium chloride, in any of three types of cells: diaphragm, mercury, or membrane. The majority of chlorine produced in the United States is manufactured by the electrolysis of sodium chloride to form chlorine gas and sodium hydroxide in diaphragm cells. The mercury cell process produces a more concentrated caustic solution (50%) than the diaphragm cell. Chlorine gas can also be generated by the salt process (which employs the reaction between sodium chloride and nitric acid), by the hydrochloric acid oxidation process, and by the electrolysis of hydrochloric acid solutions. The gas is shipped under pressure in 150-lb cylinders, 1-ton cylinders, tank trucks, tank cars, and barges.

Appendix 3a

Guidelines for Drinking Water Quality

Environmental and Workplace Health: Guidelines for Canadian
Drinking Water Quality - Summary Table
http://www.hc-sc.gc.ca/ewh-semt/pubs/water-eau/2010-sum_guide-
res_recom/index-eng.php#a4

Table 1. New and revised guidelines

Parameter	Guideline (mg/L)	Previous guideline (mg/L)	CHE approval
Microbiological parameters[a]			
Bacteriological		0 coliforms/100 mL	
E.coli	0 per 100 mL		2006
Total coliforms	0 per 100 mL		2006
Heterotrophic plate count	No numerical guideline required		2006
Emerging pathogens	No numerical guideline required		2006
Protozoa	No numerical guideline required	None	2004
Enteric viruses	No numerical guideline required	None	2004
Turbidity	0.3/1.0/0.1 NTU[b]	1.0 NTU	2004
Chemical and physical parameters			
Aluminum	0.1/0.2[c]	None	1999
Antimony	0.006	None	1997
Arsenic	0.010	0.025	2006
Benzene	0.005	0.005	2009
Bromate	0.01	None	1999
Chlorate	1.0	None	2008
Chlorine	No numerical guideline required	None	2009

[a]Refer to section on Guidelines for microbiological parameters.

[b]Based on conventional treatment/slow sand or diatomaceous earth filtration/membrane filtration.

[c]This is an operational guidance value, designed to apply only to drinking water treatment plants using aluminum-based coagulants. The operational guidance values of 0.1 mg/L applies to conventional treatment plants, and 0.2 mg/L applies to other types of treatment systems.

[d]The separate guideline for BDCM was rescinded based on new science. See addendum to the THM document. In certain situations, the Federal-Provincial-Territorial Committee on Drinking Water may choose to develop guidance documents: for contaminants that do not meet the criteria for guideline development, and for specific issues for which operational or management guidance is warranted.

Appendix 3b

Guidelines for Drinking Water Quality

Parameter	Guideline (mg/L)	Previous guideline (mg/L)	CHE approval
Chlorite	1.0	None	2008
Cyanobacterial toxins--microcystin-LR	0.0015	None	2002
Fluoride	1.5	1.5	1996
Formaldehyde	No numerical guideline required	None	1998
Haloacetic Acids--Total (HAAs)	0.080	None	2008
2-Methyl-4-chlorophenoxyacetic acid (MCPA)	0.1	None	2010
Methyl *tertiary*-butyl ether (MTBE)	0.015	None	2006
Trichloroethylene (TCE)	0.005	0.05	2005
Trihalomethanes--Total (THMs)[d]	0.100	0.100	2006
Uranium	0.02	0.1	2000
Radiological parameters			
Cesium-137 (^{137}Cs)	10 Bq/L	10 Bq/L	2009
Iodine-131 (^{131}I)	6 Bq/L	6 Bq/L	2009
Lead-210 (^{210}Pb)	0.2 Bq/L	0.1 Bq/L	2009
Radium-226 (^{226}Ra)	0.5 Bq/L	0.6 Bq/L	2009
Strontium-90 (^{90}Sr)	5 Bq/L	5 Bq/L	2009
Tritium (^{3}H)	7000 Bq/L	7000 Bq/L	2009

[a]Refer to section on Guidelines for microbiological parameters.

[b]Based on conventional treatment/slow sand or diatomaceous earth filtration/membrane filtration.

[c]This is an operational guidance value, designed to apply only to drinking water treatment plants using aluminum-based coagulants. The operational guidance values of 0.1 mg/L applies to conventional treatment plants, and 0.2 mg/L applies to other types of treatment systems.

[d]The separate guideline for BDCM was rescinded based on new science. See addendum to the THM document. In certain situations, the Federal-Provincial-Territorial Committee on Drinking Water may choose to develop guidance documents: for contaminants that do not meet the criteria for guideline development, and for specific issues for which operational or management guidance is warranted.

Chapter 3 - Chlorination is Important

As cited, "chlorine is a disinfectant added to drinking water to reduce or eliminate microorganisms, such as bacteria and viruses, which can be present in water supplies. The addition of chlorine to our drinking water has greatly reduced the risk of waterborne diseases."[1]

"Chlorine is added as part of the drinking water treatment process. However, chlorine also reacts with the organic matter, naturally present in water, such as decaying leaves. This chemical reaction forms a group of chemicals known as disinfection by-products."[2]

"The most common of these by-products are trihalomethanes (THMs), which include chloroform. The amount of THMs found in drinking water depends on a number of things, including the season and the source of the water."[3]

"For example, THM levels are generally lower in winter than in summer, because the amount of natural organic matter is lower and less chlorine is needed to disinfect at colder temperatures."[4]

"Lab animals exposed to very high levels of THMs have an increased risk of cancer. Several studies on humans have also found a link between long-term exposure to high levels of chlorination by-products and a higher risk of cancer. For instance, a recent study showed an increased risk of bladder and possibly colon cancer in people who drank chlorinated water for 35 years or more."[5]

"High levels of THMs may also have an effect on pregnancy. A California study found that pregnant women who drank large amounts of tap water with high THMs had an increased risk of miscarriage. These studies do not prove that there is a link between THMs and cancer or miscarriage. However, they do show the need for further research in this area to confirm potential health effects."[6]

"Health Canada has established a guideline for THMs of 0.1 milligrams per litre. The cancer risk at this level over a lifetime is considered extremely low. The guidelines for THMs and other chlorination by-products are currently under review by a task group whose work is coordinated by Health Canada."[7]

"Current scientific data shows that the benefits of chlorinating our drinking water (less disease) are much greater than any health risks from THMs and other by-products."[8]

"Although other disinfectants are available, chlorine remains the choice of water treatment experts. When used with modern water filtration methods, chlorine is effective against virtually all microorganisms. Chlorine is easy to apply and small amounts of the chemical remain in the water as it travels in the distribution system from the treatment plant to the consumer's tap."[9]

"This level of effectiveness ensures that microorganisms cannot re-contaminate the water after it leaves the treatment plant."[10]

Footnotes

1 - 10. *Drinking Water Chlorination*
http://www.hc-sc.gc.ca/hl-vs/iyh-vsv/environ/chlor-eng.php

Chapter 4 - Water Testing at Home

As noted, "water testing is an inexpensive and effective way of determining the quality of home drinking water."[1]

"Many home drinking water test kits have become available to the consumer in recent years and can be found in local hardware stores, health food stores and supermarkets across Canada."[2]

"Many of these tests are simple cardboard strip tests that, after being exposed to water, inform the user of the level of contamination by color change."[3]

"Home tests are now available for bacteria, lead, pesticides, nitrates, nitrites, arsenic, chlorine, pH and hardness and can be purchased as individual tests or "all in one" test kits that include combinations of tests for any or all of the above contaminants."[4]

"Here are brief descriptions of some major contaminants and their effects"[5]:

"Bacteria

*Toxic bacteria may enter the water supply from human or animal wastes or natural sources. Multiplying rapidly, they may release a variety of potent, damage causing molecules called endotoxins. Many strains of bacteria are not toxic, but some can cause very serious illness. Even mild cases can result in diarrhea, vomiting, cramps, and other gastrointestinal symptoms. Since contaminated water may not taste or smell bad, most cases of water-borne disease are not likely to be identified as such."[6]

"Lead

*Since lead was used extensively in plumbing until the 80's, many homes and buildings have pipes and plumbing fixtures that contain lead. Lead can leach from pipes into household water, making this plumbing a major source of toxic lead poisoning. Lead poisoning is so toxic that even very low levels can be dangerous. Lead consumption and poisoning has been linked to learning disabilities, muscle and bone disorders and kidney damage. Lead levels should be tested at each faucet in the home, especially if plumbing fixtures could date from the 80's or before."[7]

"Pesticides

Pesticides are chemicals used to eliminate weeds, insects and other harmful elements in crops. Atrazine and simazine are two of the pesticides most commonly found to contaminate drinking water. Millions of pounds of these two chemicals are introduced into the environment each year as herbicides and left to potentially leak into the soil, and then into groundwater, lakes and rivers that are sources of drinking water. These chemicals are so toxic that the USEPA-mandated maximum contaminant level in drinking water is equivalent to less than one drop in a large swimming pool." [8]

"Chlorine

Chlorine is used extensively to purify water. The consumption of chlorine in very small amounts will likely not cause serious harm, but its by-products, including chloroform, which are produced when chlorine mixes with organic matter, can be extremely harmful. Drinking water with high levels of chlorine by-products has been linked to some forms of cancer." [9]

"Nitrates & Nitrites

When animal and human wastes or field fertilizers come into contact with water, they produce nitrates and nitrites. Both are dangerous to children and can cause "Blue Baby Syndrome," a lethal form of birth defect in infants." [10]

"pH and Hardness

PH and Hardness do not directly cause harm but can cause a variety of secondary effects if not treated. If water acidity is too high, corrosion can leach out lead from pipes and plumbing (see lead above), as well as damage plumbing and water heating systems. Water hardness is primarily caused by calcium and magnesium compounds. These chemicals are not easily detected, but the negative effects include scaling of pots and pans and, if left untreated, damage to plumbing and water heaters." [11]

With respect to Atrazine, it was cited that "Since 1959, US farmers have counted on the herbicide atrazine to control weeds in corn, grain sorghum, sugar cane and other crops. It remains a favorite after 50 years because it works better and costs less than most alternatives. Atrazine still stands up to the most stringent safety tests and regulatory standards in the world—those of the US Environmental Protection Agency."[12]

"Despite this, Holiday Shores Sanitary District has filed suit against Syngenta and other atrazine manufacturers and distributors in Madison County, Illinois, claiming exposure to atrazine in water at any measurable level could pose a risk to property values and people."[13]

"The reality is that atrazine, according to the US EPA, "is one of the most closely examined pesticides in the marketplace" and is strictly regulated in water. After a stringent 11-year review-and the weight of evidence of more than 5,000 studies-EPA recently re-registered atrazine for use in agriculture."[14]

However, in Europe they have banned atrazine.

As cited, "it is often claimed that atrazine is of great economic benefit to corn growers, but support for this claim is limited. Some cost–benefit studies have assumed that atrazine boosts corn yields by 6%; an extensive review found a 3%–4% average yield increase; other research suggests only a 1% yield effect."[15]

"Regulation of pesticides has followed a different path in Europe than in the United States, with important implications for atrazine. The divergence dates back at least to the European Union's 1980 Drinking Water Directive, which specified 5 µg/L as the maximum allowable level of any pesticide in drinking water."[16]

By 1998 the allowable limit had been lowered to 0.1 µg/L of any one pesticide and no more than 0.5 µg/L of total pesticides. Meanwhile, a 1991 EU directive on pesticides curtailed the use of products suspected of harming human health, groundwater, or the environment. It also established a 12-year review period for products already on the market, such as atrazine, to determine their impacts."[17]

"Twelve years later, in 2003, the scientific committee reviewing atrazine concluded that it had the potential to contaminate groundwater at levels exceeding the allowed 0.1 µg/L even when used appropriately. This set in motion the process for a regulatory ban. In 2004 the Commission announced a ban on atrazine applying to all EU member states, which went into effect in 2005."[18]

With respect to simazine ($C_7H_{12}ClN_5$), Canada reports that it is "a triazine soil sterilant and pre-emergence herbicide used in the control of broadleaf and grassy weeds for a wide variety of crops."[19]

"Between 100 000 and 500 000 kg are used annually in Canada."[20]

"The solubility of simazine in water at 20°C is 3.5 mg/L; its vapour pressure is 8.1 × 10-7 Pa at 20°C."[21]

"The log octanol-water partition coefficient of simazine is reported to be 1.94; it is therefore not likely to bioaccumulate to a significant degree in human or animal tissue."[22]

As further cited, "simazine was detected in nine of 440 surface water samples (mean detectable concentration 0.6 µg/L) from three Ontario river basins surveyed from 1981 to 1985 (detection limit 0.2 µg/L); a total of only 800 kg had been used in these areas in 1983."[23]

And, "simazine was detected in 55 of 1199 samples of municipal and private drinking water supplies in Nova Scotia (1986), Quebec (1986), Ontario (1979 to 1986), Manitoba (1986) and Alberta (1978 to 1986) (detection limits ranged from 0.025 to 1.0 µg/L)."[24]

"Conventional treatment processes are reported to be relatively ineffective in removing simazine from drinking water supplies."[25]

In Europe, it's a different story?

"Simazine was already banned in Norway, Belgium and France. This ban will now be extended to all member states of the EU."[26]

"Simazine in water may be determined by extraction with chloroform, separation by gas/liquid chromatography and detection by electrolytic conduction, nitrogen mode (detection limit 0.2 µg/L)."[27]

"An alternative method involves extraction into dichloromethane, gas chromatographic capillary column separation and nitrogen phosphorus detection (estimated detection limit 0.1 to 2 µg/L)."[28]

Footnotes

1 - 11. *Is Your Drinking Water Safe?*
A Guide to Water Testing
Brett Hopkins
Silver Lake Research - Watersafe
March 2002
http://www.elements.nb.ca/theme/water/silverlake/testing.htm

12 - 14. *Welcome to AtrazineFacts.com*
http://www.atrazinefacts.com/

15 - 18. *The Economics of Atrazine*
Frank Ackerman, Phd
http://docs.google.com/viewer?a=v&q=cache:kJXlYvfhnkAJ:ase.tufts.
edu/gdae/Pubs/rp/EconAtrazine.pdf+The+European+Community+has
+banned+Atrazine&hl=en&gl=ca&pid=bl&srcid=ADGEESgALGRE3
PmGaOWPWV0tNRrQlB_KblIVwvQ-
lvPPJaYxRo33Nx6T_a4tJrD9AZjf55tfr5SOcMI5dxXsJptBH4hglUc9
AWtCBChGRvQI2KmYrb8KhkIvaDTX6YagilqypIIXne1N&sig=AHI
EtbQgFbosCQ23GuQLJgTd-MoVcq_BZg

Also see: *European Council. Council Directive 80/778/EEC of 15 July
1980 relating to the quality of water intended for human consumption.*
Official Journal of the European Union. 1980; L229:11-29.

Also see: *European Council. Council Directive 98/83/EC of 3
November 1998 on the quality of water intended for human
consumption.* Official Journal of the European Union. 1998;L330:32-
54.

Also see: *European Council. Council Directive of 15 July 1991
concerning the placing of plant protection products on the market,*
91/414/EEC. Official Journal of the European Communities.
1991; L230:1-40.

19 - 20. *Simazine*
http://www.hc-sc.gc.ca/ewh-semt/pubs/water-eau/simazine/index-
eng.php

Also see: *Environment Canada/Agriculture Canada. Pesticide
Registrant Survey, 1986 report.* Commercial Chemicals Branch,
Conservation and Protection, Environment Canada, Ottawa (1987).

21. *Simazine*
http://www.hc-sc.gc.ca/ewh-semt/pubs/water-eau/simazine/index-
eng.php

Also see: *Agriculture Canada. Guide to the chemicals used in crop
protection. 7th edition.* Publication No. 1093 (1982).

22. *Simazine*
http://www.hc-sc.gc.ca/ewh-semt/pubs/water-eau/simazine/index-
eng.php

Also see: Suntio, L.R., Shiu, W.Y., Mackay, D., Seiber, J.N. and Glotfelty, D. *Critical review of Henry's Law constants for pesticides*. Rev. Environ. Contam. Toxicol., 103: 1 (1988).

23. *Simazine*
http://www.hc-sc.gc.ca/ewh-semt/pubs/water-eau/simazine/index-eng.php

Also see: Frank, R. and Logan, L. *Pesticide and industrial chemical residues at the mouth of the Grand, Saugeen and Thames rivers, Ontario, Canada, 1981-85*. Arch. Environ. Contam. Toxicol., 17: 741 (1988).

24 - 25. *Simazine*
http://www.hc-sc.gc.ca/ewh-semt/pubs/water-eau/simazine/index-eng.php

Also see: Department of National Health and Welfare. *National pesticide residue limits in food*. Food Directorate, Ottawa (1986).

26. *Atrazine and simazine banned in the EU but 'essential uses' remain in the UK*
http://www.pan-uk.org/pestnews/Issue/pn62/pn62p19a.htm

Also see: *Basic Information about Simazine in Drinking Water*
http://water.epa.gov/drink/contaminants/basicinformation/simazine.cfm

27. *Simazine*
http://www.hc-sc.gc.ca/ewh-semt/pubs/water-eau/simazine/index-eng.php

Also see: Frank, R. and Logan, L. *Pesticide and industrial chemical residues at the mouth of the Grand, Saugeen and Thames rivers, Ontario, Canada, 1981-85*. Arch. Environ. Contam. Toxicol., 17: 741 (1988).

28. *Simazine*
http://www.hc-sc.gc.ca/ewh-semt/pubs/water-eau/simazine/index-eng.php

Also see: U.S. Environmental Protection Agency. *Health advisory -- simazine*. Office of Drinking Water (1987).

Chapter 5 - Presence – Absence Tests

What this experiment is interested in is a quick – simple test is for coliform bacteria in the drinking – tap water in Winnipeg, Manitoba, Canada.

Where, "coliform bacteria are described and grouped, based on their common origin or characteristics, as either Total or Fecal Coliform."[1]

"The Total group includes Fecal Coliform bacteria such as Escherichia coli (E .coli), as well as other types of Coliform bacteria that are naturally found in the soil."[2]

"Fecal Coliform bacteria exist in the intestines of warm blooded animals and humans, and are found in bodily waste, animal droppings, and naturally in soil. Most of the Fecal Coliform in fecal material (feces) is comprised of E. coli, and the serotype E. coli 0157:H7 is known to cause serious human illness."[3]

"Total Coliform do not necessarily indicate recent water contamination by fecal waste, however, the presence or absence of these bacteria in treated water is often used to determine whether water disinfection is working properly."[4]

An appropriate method to test, "a basic laboratory test is the best way to tell if Coliform organisms are present, as they can be there with no appearance or taste difference."[5]

"When water is tested for Fecal or Total Coliform, the results are usually given as the number of colony forming units per 100 millilitres (CFU/100ml) of water sampled"[6] [see **Appendix 4**, cited below].

"No sample should contain Fecal Coliform or E. coli, and ideally there should be no Total Coliform, however a single sample may contain up to 10 Total Coliform CFU/100 ml"[7] [see **Appendix 5**, cited below].

Investigating the difference, it was noted "β-galactosidase and β-glucuronidase-based commercial culture methods used to assess water quality, "their analytical performance, in terms of their respective ability to detect different strains of Escherichia coli and total coliforms, had never been systematically compared with pure cultures."[8]

β-glucuronidase production

"Here, their ability to detect β-glucuronidase production from E. coli isolates was evaluated by using 74 E. coli strains of different geographic origins and serotypes encountered in fecal and environmental settings."[9]

"Chromocult, MI, Readycult, and Colilert detected β-glucuronidase production from respectively 79.9, 79.9, 81.1, and 51.4% of the 74 E. coli strains tested."[10]

β-galactosidase production

"Their ability to detect β-galactosidase production was studied by testing the 74 E. coli strains as well as 33 reference and environmental non-E. coli total coliform strains."[11]

Chromocult, MI, Readycult, and Colilert "detected β-galactosidase production from respectively 85.1, 73.8, 84.1, and 84.1% of the total coliform strains tested."[12]

"The results of the present study suggest that Colilert is the weakest method tested to detect β-glucuronidase production and MI the weakest to detect β-galactosidase production."[13]

"Furthermore, the high level of false-negative results for E. coli recognition obtained by all four methods suggests that they may not be appropriate for identification of presumptive E. coli strains."[14]

In addition, "total coliforms are not necessarily an indication of the presence of faecal contamination."[15]

"Faecal coliforms in drinking water may, however, indicate the presence of faecal contamination."[16]

"The presence of Escherichia coli, one species in the faecal coliform group, is a definite indicator of the presence of faeces."[17]

"Other species in the faecal coliform group (e.g., Klebsiella pneumoniae, Enterobacter cloacae) are not restricted to faeces but occur naturally on vegetation and in soils."[18]

"The MAC for coliforms in drinking water is zero organisms detectable per 100 mL."[19]

"Because coliforms are not uniformly distributed in water and are subject to considerable variation in enumeration, drinking water that fulfils the following conditions is considered to be in compliance with the coliform MAC:

1. No sample should contain more than 10 total coliform organisms per 100 mL, none of which should be faecal coliforms;

2. No consecutive sample from the same site should show the presence of coliform organisms; and

3. For community drinking water distribution systems:

a) not more than one sample from a set of samples taken from the community on a given day should show the presence of coliform organisms; and

b) not more than 10% of the samples based on a minimum of 10 samples should show the presence of coliform organisms."[20]

The authors wanted a quick test of their tap water in Winnipeg, Manitoba, Canada and chose LaMotte's Urban Water Test Kit[21], a simple test described as for "grades 3 and up [**Figure 1**, appearing below].

Figure 1: Urban Water Quality Test Kit 5918

-

As described in the write-up for this simple Presence - Absence tap water test kit, "the life style of youth in the city often does not include involvement in outdoor activities or contact with nature."[22]

"They may feel that environmental issues are not a part of their world when actually, urban environmental issues involving waters are all around them."[23]

"This kit provides a hands-on introduction to the basic concepts of water chemistry and environmental awareness in urban areas."[24]

"All of the necessary equipment is included to test for the following test factors using our non-hazardous Testab methods - Chlorine, Copper, Iron, Hardness, Nitrate, pH, Phosphates, and Temperature."[25]

"Recommended for ages 6 and up."[26]

What could be simpler?

Footnotes

1 - 7. *WATER STEWARDSHIP INFORMATION SERIES: Total, Fecal & E. coli Bacteria in Groundwater*
http://docs.google.com/viewer?a=v&q=cache:9dtVokUmtAwJ:www.env.gov.bc.ca/wsd/plan_protect_sustain/groundwater/library/ground_fact_sheets/pdfs/coliform(020715)_fin2.pdf+total+coliform&hl=en&gl=ca&pid=bl&srcid=ADGEEShUFl7d_kYfnx2nX_aCmDWLiA2OLbBp1BgFpLs8-oKlDz1SGnm8N3yXVGdtL6Yy1ApbK-0OchYxtwtzlFeDs3_LAHdUczNuYOJTm011QhN-6415dPfnaN-53h6pimdLWEx8PdT0&sig=AHIEtbT6Q1zQ8T48q1lX5E6tJV3s_hOb4w

8 - 14. Andrée F. Maheux, Vicky Huppé, Maurice Boissinot, François J. Picard, Luc Bissonnette, Jean-Luc T. Bernier, Michel G. Bergeron, *Analytical limits of four β-glucuronidase and β-galactosidase-based commercial culture methods used to detect Escherichia coli and total coliforms.* Journal of Microbiological Methods 75 (2008) 506–514.
http://docs.google.com/viewer?a=v&q=cache:fgcMaiMMuqEJ:www.atlantis.ulaval.ca/publications/fichiers/Culture%2520Methods.pdf+WATER+%CE%B2-glucuronidase&hl=en&gl=ca&pid=bl&srcid=ADGEESg1fjbxmoFWun_Qbar-5AxuLxVpyOwpnRNyS6i1cB-aQPxqUAV-iRKQTn3IfhMqdoYQEKCCSmaW3byHIJhV-7tpiTAbA0juhjRR6scf7-HHDbpGKEAlmrGwvgdgirMkla-vQHRJ&sig=AHIEtbQZYx4McCcDoDZL9oadzskojxGN4g

Also see: Ley, A.N., Bowers, R.J. and Wolfe, S. (1988). *Indoxyl-beta-D-glucuronide, a novel chromogenic reagent for the specific detection and enumeration of Escherichia coli in environmental samples.* Can. J. Microbiol. 34: 690-693.
http://www.ncbi.nlm.nih.gov/pubmed/3061623

See also: L S Warren, R E Benoit, and J A Jessee, *Rapid enumeration of Fecal Coliforms in water by a colorimetric beta-galactosidase assay.* Appl Environ Microbiol. 1978 January; 35(1): 136–141. PMCID: PMC242792
http://www.ncbi.nlm.nih.gov/pmc/articles/PMC242792/

For Chromocult, see also: Dennis Byamukama, Frank Kansiime, Robert L. Mach, and Andreas H. Farnleitner, *Determination of Escherichia coli Contamination with Chromocult Coliform Agar Showed a High Level of Discrimination Efficiency for Differing Fecal Pollution Levels in Tropical Waters of Kampala, Uganda,* Appl Environ Microbiol. 2000 February; 66(2): 864–868. PMCID: PMC91912. American Society for Microbiology.
http://www.ncbi.nlm.nih.gov/pmc/articles/PMC91912/

For MI, see also: *Method 1604: Total Coliforms and Escherichia coli in Water by Membrane Filtration Using a Simultaneous Detection Technique (MI Medium).* EPA-821-R-02-024, U.S. Environmental Protection Agency, Office of Water (4303T), Washington, DC 20460.
http://docs.google.com/viewer?a=v&q=cache:CAOLAJAKp3IJ:www.epa.gov/microbes/1604sp02.pdf+MI+coliform&hl=en&gl=ca&pid=bl&srcid=ADGEESjrQ2X25NFfc8TEbYhp8joFL7eFWav5t8VjhYc-Sn_i2CaJhyIGuH4swis99xfYpbXtDM0MINjjyytOgiGiKYBGcXfc2KIb9P1FBQqDqH_ko7iDhfwnGXt_CLVHeXDP_o8wLOcv&sig=AHIEtbRYx7jhS8JX8scLbI5ONykMyAUzmw

For Readycult, see also: *Readycult® Coliforms 100 Presence/Absence Test for Detection and Identification of Coliform Bacteria and Escherichia coli in Finished Waters.* January 2007, Version 1.1, EPA-approved alternative version for November 2000, Version 1.0. MERCK KGaA, Germany.
http://docs.google.com/viewer?a=v&q=cache:UMO0iitRol8J:www.univie.ac.at/hygiene-aktuell/RC.pdf+readycult+coliforms&hl=en&gl=ca&pid=bl&srcid=ADGEESgzqJzobif6NfLg8ero_uk_xqe1CI9ws_FbRhC0Oq8ZBKjjCO8iaotDIET-Jw5LIiHQJiQ25Bko3xMyw9rycdesr0hGOpZL-FRjmXy2M_jUShWIxMIVrQXALbr34Bo08kEGYKbK&sig=AHIEtbTeK0oydShTIDmO9EzcxkffdOK2tA

For Colilert, see also: Kuo-Kuang Chao, Chen-Ching Chao and Wei-Liang Chao, *Evaluation of Colilert-18 for Detection of Coliforms and Eschericha coli in Subtropical Freshwater.* Appl. Environ. Microbiol. February 2004 vol. 70 no. 2 1242-1244.
http://aem.asm.org/content/70/2/1242.full

15 - 20. *Canada Water Act, Guidelines for Canadian Drinking Water Quality.* Last updated, 03/31/2011
http://www.voccompliance.com/elis/elis_docs.asp?doc_id=e05

21 - 26. LaMotte's Urban Water Test Kit
http://www.lamotte.com/component/option,com_pages/page,88/task,item/

Appendix 4

How do you convert mg/L (milligrams per liter) to ppm (parts per million)?

A : We can use them interchangeably for well water measurements: 1mg/L = 1ppm.

How do I convert gpg (grains per gallon) into mg/L (milligrams per liter)?

A : Generally, mg/L is used by the scientific community and gpg is used by the water treatment industry. The conversion formula is: 1 gpg = 17.1 mg/L.

Reference to: *State Hygienic Laboratory at The University of Iowa*
http://shl.uiowa.edu/services/wellwater/faq.xml

Appendix 5

Guidelines for Drinking Water Quality

Environmental and Workplace Health: Guidelines for Canadian Drinking Water Quality - Summary Table
http://www.hc-sc.gc.ca/ewh-semt/pubs/water-eau/2010-sum_guide-res_recom/index-eng.php#a4

Table 1. New and revised guidelines

Parameter	Guideline (mg/L)	Previous guideline (mg/L)	CHE approval
Microbiological parameters[a]			
Bacteriological		0 coliforms/100 mL	
E.coli	0 per 100 mL		2006
Total coliforms	0 per 100 mL		2006
Heterotrophic plate count	No numerical guideline required		2006
Emerging pathogens	No numerical guideline required		2006
Protozoa	No numerical guideline required	None	2004
Enteric viruses	No numerical guideline required	None	2004
Turbidity	0.3/1.0/0.1 NTU[b]	1.0 NTU	2004
Chemical and physical parameters			
Aluminum	0.1/0.2[c]	None	1999
Antimony	0.006	None	1997
Arsenic	0.010	0.025	2006
Benzene	0.005	0.005	2009
Bromate	0.01	None	1999
Chlorate	1.0	None	2008
Chlorine	No numerical guideline required	None	2009

[a]Refer to section on Guidelines for microbiological parameters.

[b]Based on conventional treatment/slow sand or diatomaceous earth filtration/membrane filtration.

[c]This is an operational guidance value, designed to apply only to drinking water treatment plants using aluminum-based coagulants. The operational guidance values of 0.1 mg/L applies to conventional treatment plants, and 0.2 mg/L applies to other types of treatment systems.

[d]The separate guideline for BDCM was rescinded based on new science. See addendum to the THM document. In certain situations, the Federal-Provincial-Territorial Committee on Drinking Water may choose to develop guidance documents: for contaminants that do not meet the criteria for guideline development, and for specific issues for which operational or management guidance is warranted.

Chapter 6 - Water test 07 August, 2011

Simply stated, this experiment was to test coliform levels [if any] in the tap water in Winnipeg, Manitoba, Canada on 07 August, 2011 at 1200 hours using the La Motte Urban Water Quality Test Kit 5918 [see **Figure 1**, appearing below].

Figure 1: Urban Water Quality Test Kit 5918

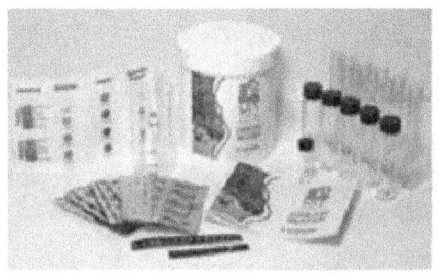

-

The test procedures are also quite simple, as described in **Figure 2**, appearing below.

Figure 2: Urban Water Quality Test Kit 5918

TEST PROCEDURES

Bacteria

Coliform bacteria are generally harmless bacteria that live naturally in the intestines of warm-blooded animals, including humans, and help the body function. Coliform bacteria are abundant in human and animal feces but do not usually occur elsewhere. Even though coliform bacteria itself may not make you sick, they are often found with other types of bacteria that are harmful. For this reason, coliform bacteria are used as an indicator of sewage or fecal contamination.

Water supplies can become contaminated with coliform bacteria when sewer lines become damaged or rainwater washes pet waste into storm sewer systems. Coliform bacteria is introduced to water in lakes, ponds and puddles by animals that live in the area, like birds and small mammals.

The sewage systems of some cities discharge sewage directly into local rivers. A person who swims in water with high levels of coliform bacteria could get sick from swallowing the water or from the bacteria entering their body through cuts or scrapes on their skin.

Even if test results are negative for coliform bacteria, water samples should be tested by a professional lab before the water is considered to be safe.

Results

As indicated in the La Motte Urban Water Quality Test Kit 5918, if there is a the Presence of Bacteria in the sample of tap water taken, there will appear "many gas bubbles present", "Gel rises to surface", "Liquid below gel is cloudy" and "Indicator turns yellow" [see **Figure 3** appearing below].

Figure 3: Urban Water Quality Test Kit 5918

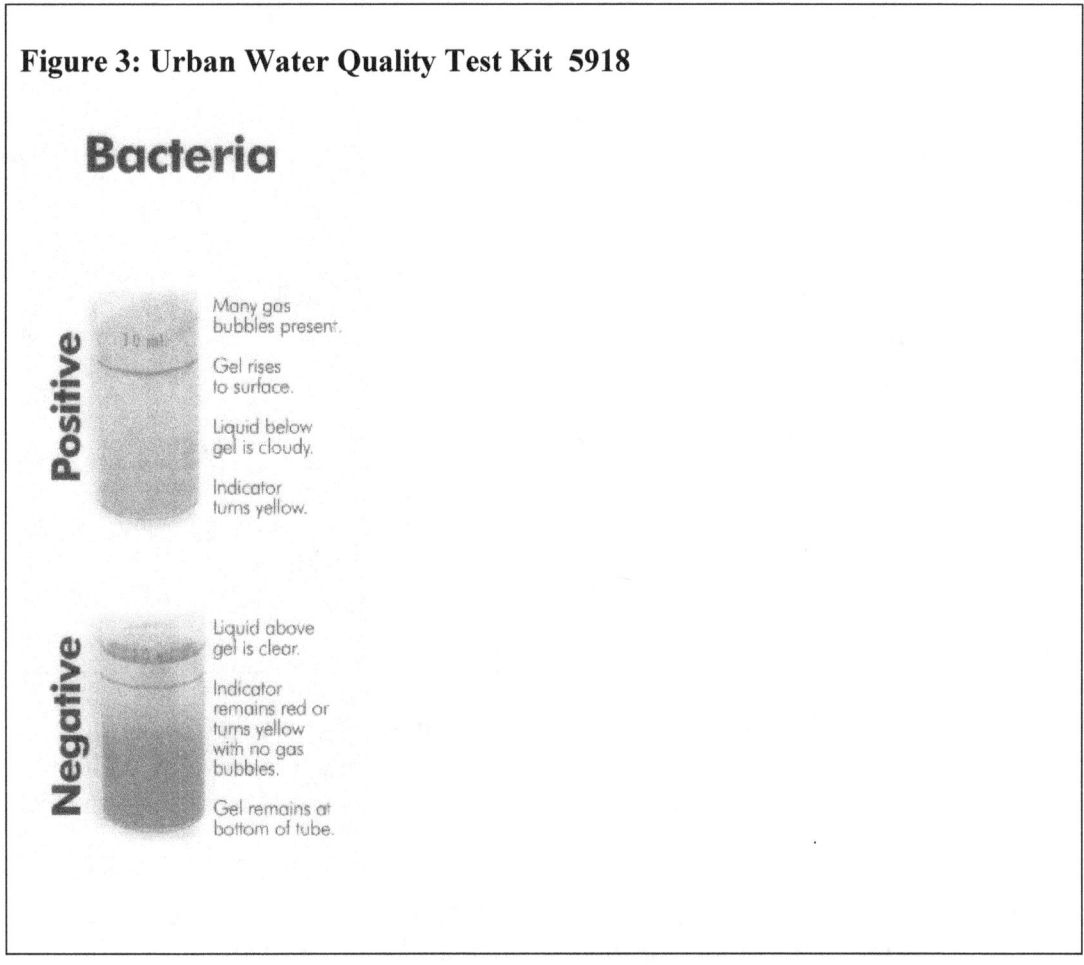

As mentioned the sample of tap water was taken in Winnipeg, Manitoba, Canada on 07 August, 2011 at 1200 hours in the eastern quadrant of the city. The weather was clear, sunny and no recent rainfall.

The tap water ran for 3 minutes before the water sample was taken.

The procedures for Bacteria testing according to La Motte Urban Water Quality Test Kit 5918 were again quite simple, as described in **Figure 4**, appearing below.

Figure 4: Urban Water Quality Test Kit 5918

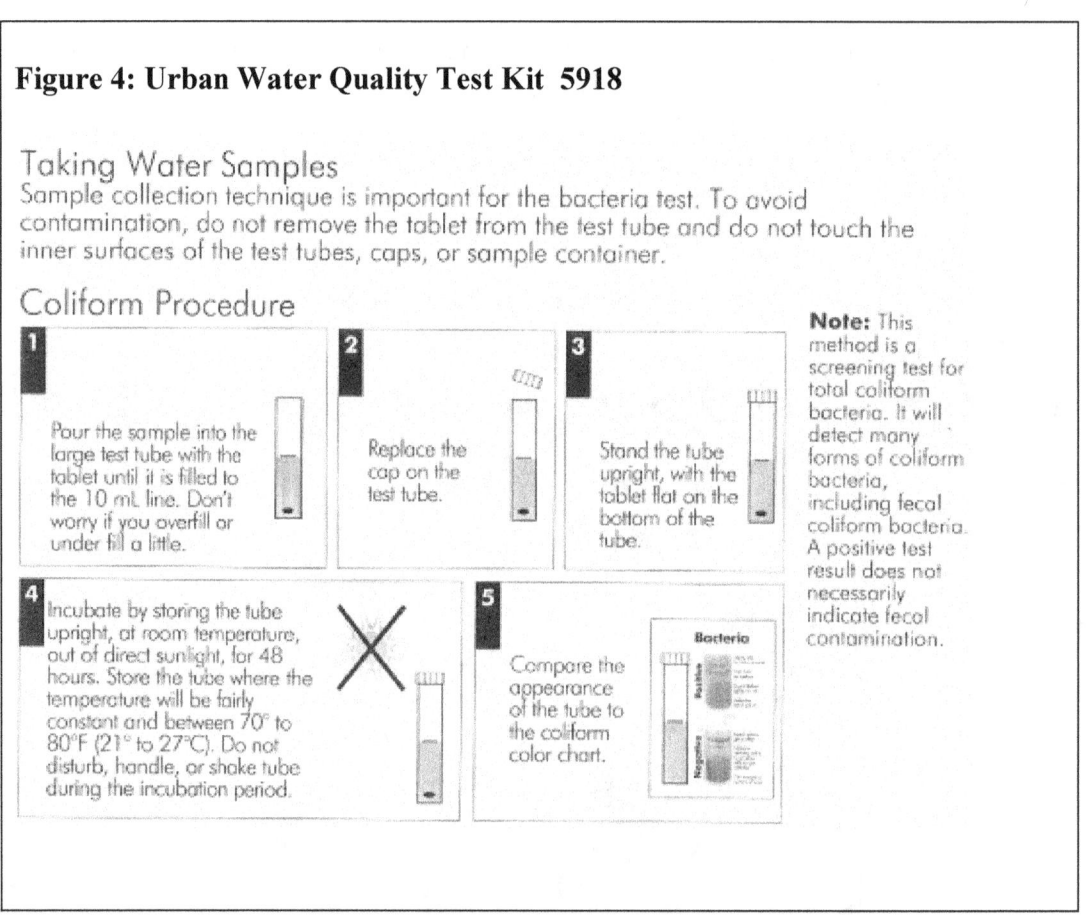

Taking Water Samples
Sample collection technique is important for the bacteria test. To avoid contamination, do not remove the tablet from the test tube and do not touch the inner surfaces of the test tubes, caps, or sample container.

Coliform Procedure

1 Pour the sample into the large test tube with the tablet until it is filled to the 10 mL line. Don't worry if you overfill or under fill a little.

2 Replace the cap on the test tube.

3 Stand the tube upright, with the tablet flat on the bottom of the tube.

4 Incubate by storing the tube upright, at room temperature, out of direct sunlight, for 48 hours. Store the tube where the temperature will be fairly constant and between 70° to 80°F (21° to 27°C). Do not disturb, handle, or shake tube during the incubation period.

5 Compare the appearance of the tube to the coliform color chart.

Note: This method is a screening test for total coliform bacteria. It will detect many forms of coliform bacteria, including fecal coliform bacteria. A positive test result does not necessarily indicate fecal contamination.

The tap water taken in Winnipeg, Manitoba, Canada on 07 August, 2011 at 1200 hours in the eastern quadrant of the city on a clear, sunny day with no recent rainfall appeared to indicate after the 48 hour waiting period that Coliform Bacteria was present.

As described in **Figure 3**, there will appear "many gas bubbles present", "Gel rises to surface", "Liquid below gel is cloudy" and "Indicator turns yellow".

That is exactly what happened [see **Figure 5** appearing below].

As the cautionary note specified in **Figure 4**, "A positive test result does not necessarily indicate fecal contamination".

Figure 5: Urban Water Quality Test Kit 5918

<u>DRINKING WATER from tap water SAMPLE TAKEN 07 AUGUST, 2011</u>

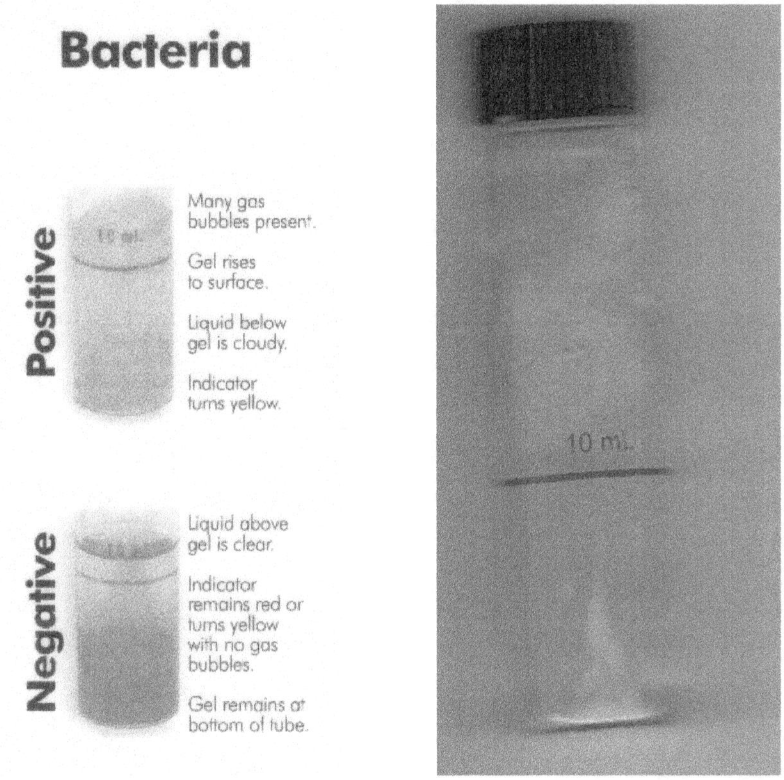

As appearing in **Figure 5**, there was "many gas bubbles present", with "Gel rises to surface", and "Liquid below gel is cloudy" as well as "Indicator turns yellow".

The tap water was, in addition to Bacteria, also tested for Chlorine, Ph, Dissolved Oxygen, Iron, Copper, Nitrate and Phosphate.

Those results were:

Chlorine – appeared to be 1 ppm
Ph – appeared to be 7
Dissolved Oxygen – appeared to be 6.5 ppm
Iron – appeared to be 0.25 ppm

Copper – appeared to be 0.0 ppm
Nitrate – appeared to be 0.0 ppm [although sample was partly cloudy]
Phosphate – appeared to be 1.75 ppm [apparently should be max of 1 ppm].

Chapter 7 - Water test 09 August, 2011

Concerned about the results that apparently indicated the presence of Bacteria Coliform in the tap water sample taken on 07 August, 2011, cited above, the authors repeated the testing on 07 August, 2011 using the same La Motte Urban Water Quality Test Kit 5918 [see **Figure 1**, appearing below].

Figure 1: Urban Water Quality Test Kit 5918

-

As mentioned, the test procedures are also quite simple, as described in **Figure 2**, appearing below.

Figure 2: Urban Water Quality Test Kit 5918

TEST PROCEDURES

Bacteria

Coliform bacteria are generally harmless bacteria that live naturally in the intestines of warm-blooded animals, including humans, and help the body function. Coliform bacteria are abundant in human and animal feces but do not usually occur elsewhere. Even though coliform bacteria itself may not make you sick, they are often found with other types of bacteria that are harmful. For this reason, coliform bacteria are used as an indicator of sewage or fecal contamination.

Water supplies can become contaminated with coliform bacteria when sewer lines become damaged or rainwater washes pet waste into storm sewer systems. Coliform bacteria is introduced to water in lakes, ponds and puddles by animals that live in the area, like birds and small mammals.

The sewage systems of some cities discharge sewage directly into local rivers. A person who swims in water with high levels of coliform bacteria could get sick from swallowing the water or from the bacteria entering their body through cuts or scrapes on their skin.

Even if test results are negative for coliform bacteria, water samples should be tested by a professional lab before the water is considered to be safe.

Results

As indicated in the La Motte Urban Water Quality Test Kit 5918, if there is a the Presence of Bacteria in the sample of tap water taken, there will appear "many gas bubbles present", "Gel rises to surface", "Liquid below gel is cloudy" and "Indicator turns yellow" [see **Figure 3** appearing below].

Figure 3: Urban Water Quality Test Kit 5918

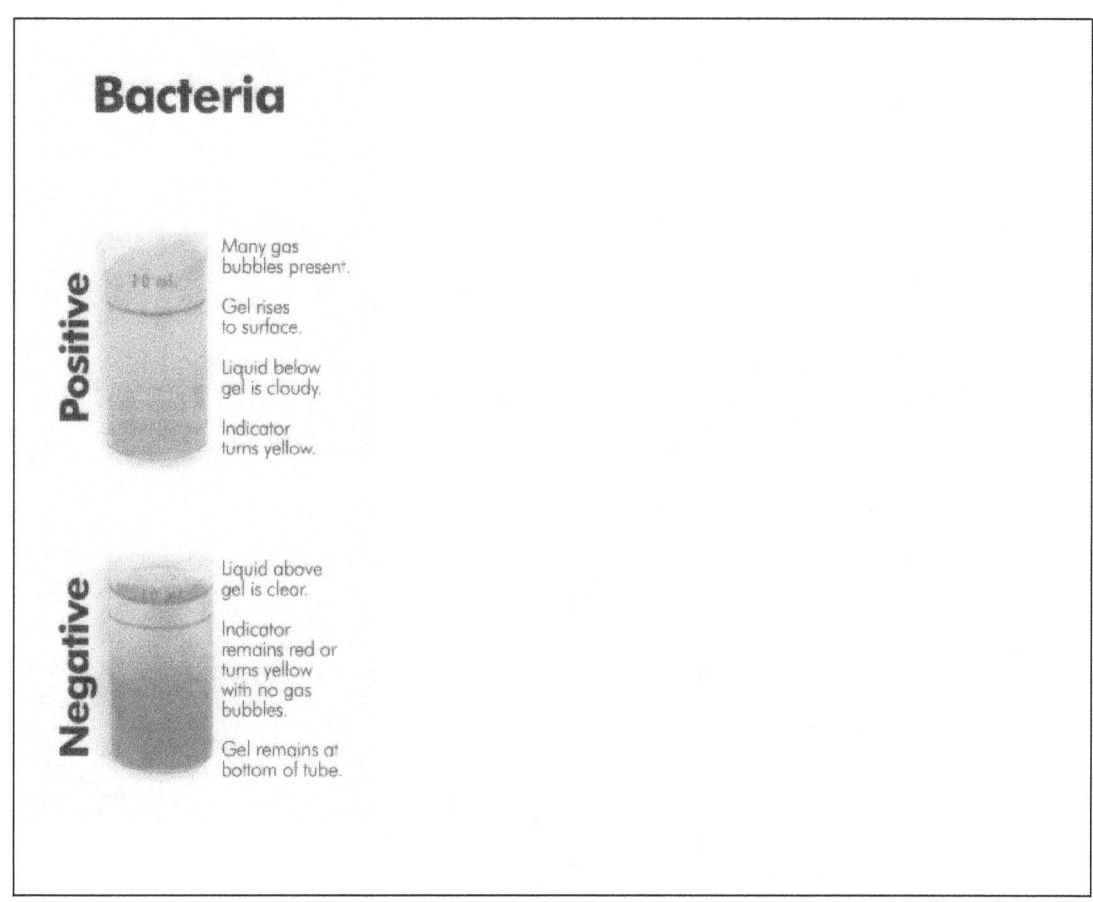

As mentioned, the procedures for Bacteria testing according to La Motte Urban Water Quality Test Kit 5918 were again quite simple, as described in **Figure 4**, appearing below.

Figure 4: Urban Water Quality Test Kit 5918

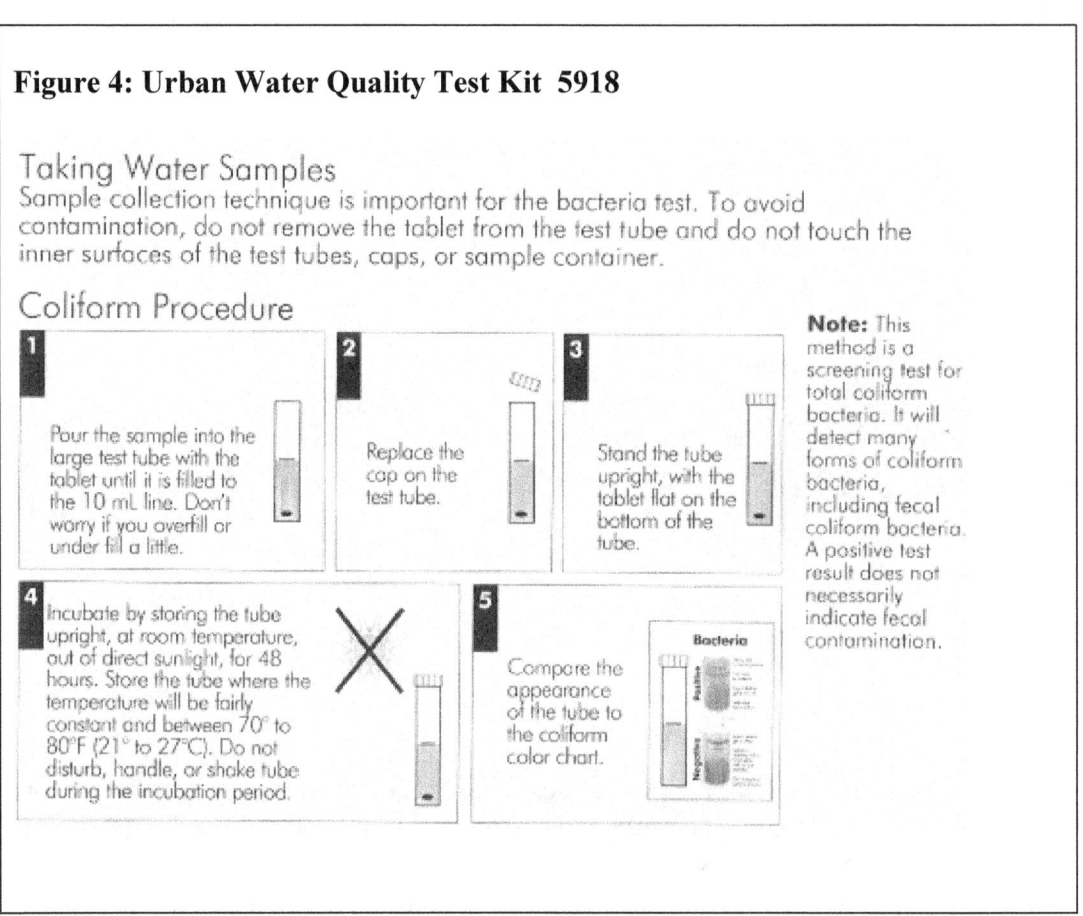

Taking Water Samples
Sample collection technique is important for the bacteria test. To avoid contamination, do not remove the tablet from the test tube and do not touch the inner surfaces of the test tubes, caps, or sample container.

Coliform Procedure

1 Pour the sample into the large test tube with the tablet until it is filled to the 10 mL line. Don't worry if you overfill or under fill a little.

2 Replace the cap on the test tube.

3 Stand the tube upright, with the tablet flat on the bottom of the tube.

4 Incubate by storing the tube upright, at room temperature, out of direct sunlight, for 48 hours. Store the tube where the temperature will be fairly constant and between 70° to 80°F (21° to 27°C). Do not disturb, handle, or shake tube during the incubation period.

5 Compare the appearance of the tube to the coliform color chart.

Bacteria

Note: This method is a screening test for total coliform bacteria. It will detect many forms of coliform bacteria, including fecal coliform bacteria. A positive test result does not necessarily indicate fecal contamination.

Again, the sample of tap water that was taken in Winnipeg, Manitoba, Canada on 09 August, 2011 at 1100 hours in the eastern quadrant of the city following a heavy rainfall on 08 August appeared to indicate after the 48 hour waiting period that Coliform Bacteria was present.

As described in **Figure 3**, there will appear "many gas bubbles present", "Gel rises to surface", "Liquid below gel is cloudy" and "Indicator turns yellow" [see **Figure 6** appearing below].

The tap water ran for 3 minutes before the water sample was taken.

That was exactly what appeared, "many gas bubbles present", with the "Gel rises to surface", and the "Liquid below gel is cloudy" as well as the "Indicator turns yellow"

Again, as the cautionary note specified in **Figure 4**, "A positive test result does not necessarily indicate fecal contamination".

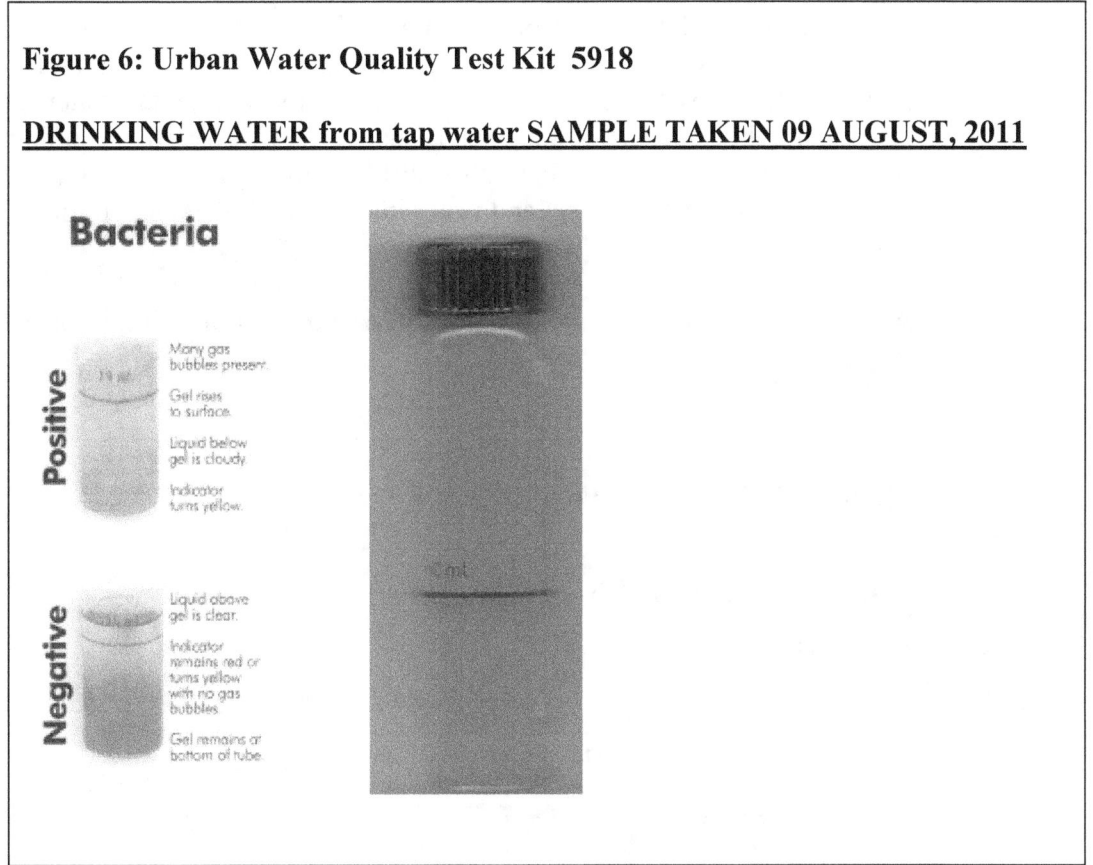

Figure 6: Urban Water Quality Test Kit 5918

DRINKING WATER from tap water SAMPLE TAKEN 09 AUGUST, 2011

The tap water was, in addition to Bacteria, also tested for Phosphate.

Those results were:

Phosphate – appeared to be 1.75 ppm [apparently should be max of 1 ppm].

Chapter 8 - Water test 06 September, 2011

Concerned about the results from 07 August, 2011 on a clear, sunny day without rainfall and then on 09 August, 2011 after a heavy rainfall on 08 August with both indicating the apparent presence of Bacteria Coliform in the tap water samples taken using the La Motte Urban Water Quality Test Kit 5918, the authors tested the water one more time on 06 September, 2011 following a heavy rain the previous days and where the tap water [and water in the toilet bowls] had turned a brownish color.

As mentioned previously, the test procedures are also quite simple, as described in **Figure 2**, appearing below.

Figure 2: Urban Water Quality Test Kit 5918

TEST PROCEDURES

Bacteria

Coliform bacteria are generally harmless bacteria that live naturally in the intestines of warm-blooded animals, including humans, and help the body function. Coliform bacteria are abundant in human and animal feces but do not usually occur elsewhere. Even though coliform bacteria itself may not make you sick, they are often found with other types of bacteria that are harmful. For this reason, coliform bacteria are used as an indicator of sewage or fecal contamination.

Water supplies can become contaminated with coliform bacteria when sewer lines become damaged or rainwater washes pet waste into storm sewer systems. Coliform bacteria is introduced to water in lakes, ponds and puddles by animals that live in the area, like birds and small mammals.

The sewage systems of some cities discharge sewage directly into local rivers. A person who swims in water with high levels of coliform bacteria could get sick from swallowing the water or from the bacteria entering their body through cuts or scrapes on their skin.

Even if test results are negative for coliform bacteria, water samples should be tested by a professional lab before the water is considered to be safe.

Results

As indicated in the La Motte Urban Water Quality Test Kit 5918, if there is a the Presence of Bacteria in the sample of tap water taken, there will appear "many gas bubbles present", "Gel rises to surface", "Liquid below gel is cloudy" and "Indicator turns yellow" [see **Figure 3** appearing below].

Figure 3: Urban Water Quality Test Kit 5918

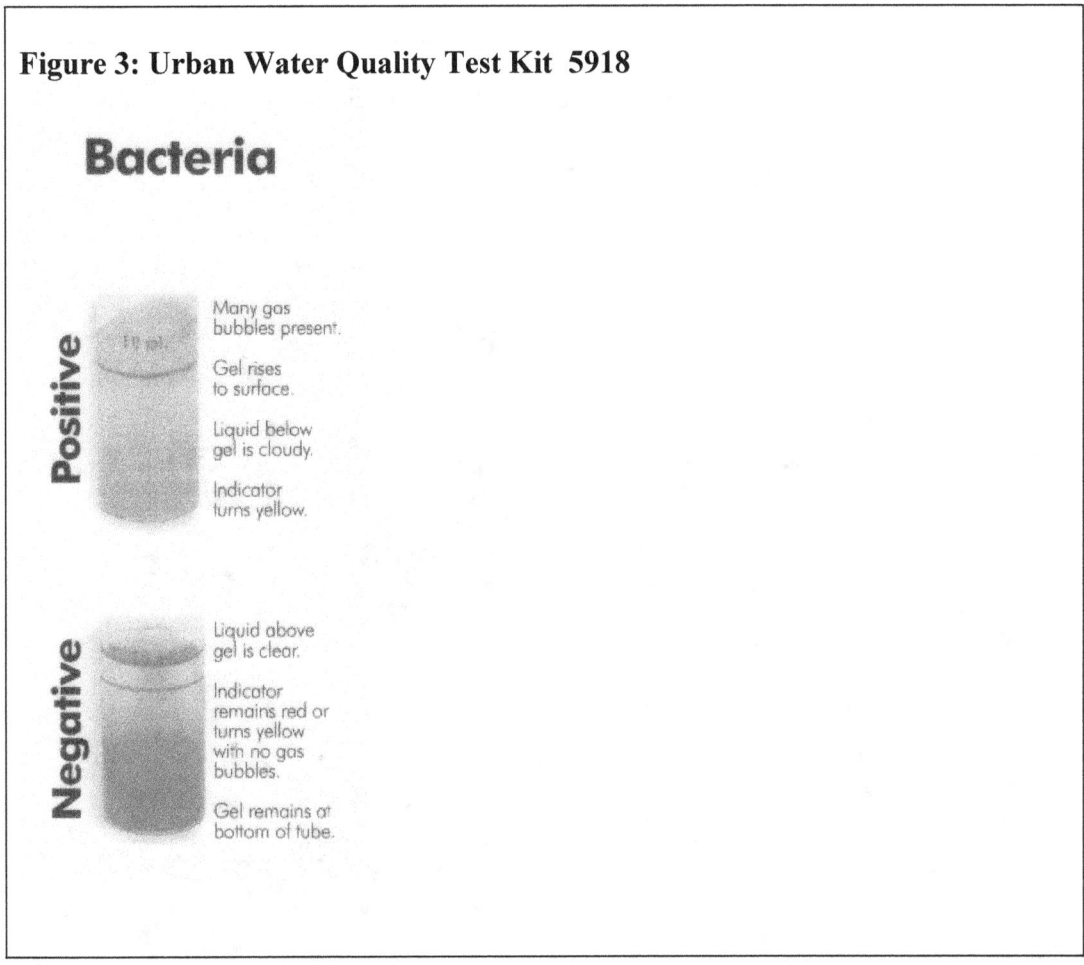

Again, the sample of tap water that was taken in Winnipeg, Manitoba, Canada on 06 September, 2011 at 10:45 hours in the eastern quadrant of the city following a heavy rainfall a few days earlier appeared to indicate after the 48 hour waiting period the presence of Coliform Bacteria, where "many gas bubbles were present", "Gel rose to surface", "Liquid below gel was cloudy" and "Indicator turned yellow" [see **Figure 7** appearing below].

Figure 7: Urban Water Quality Test Kit 5918

<u>DRINKING WATER from tap water SAMPLE TAKEN 06 SEPTEMBER, 2011</u>

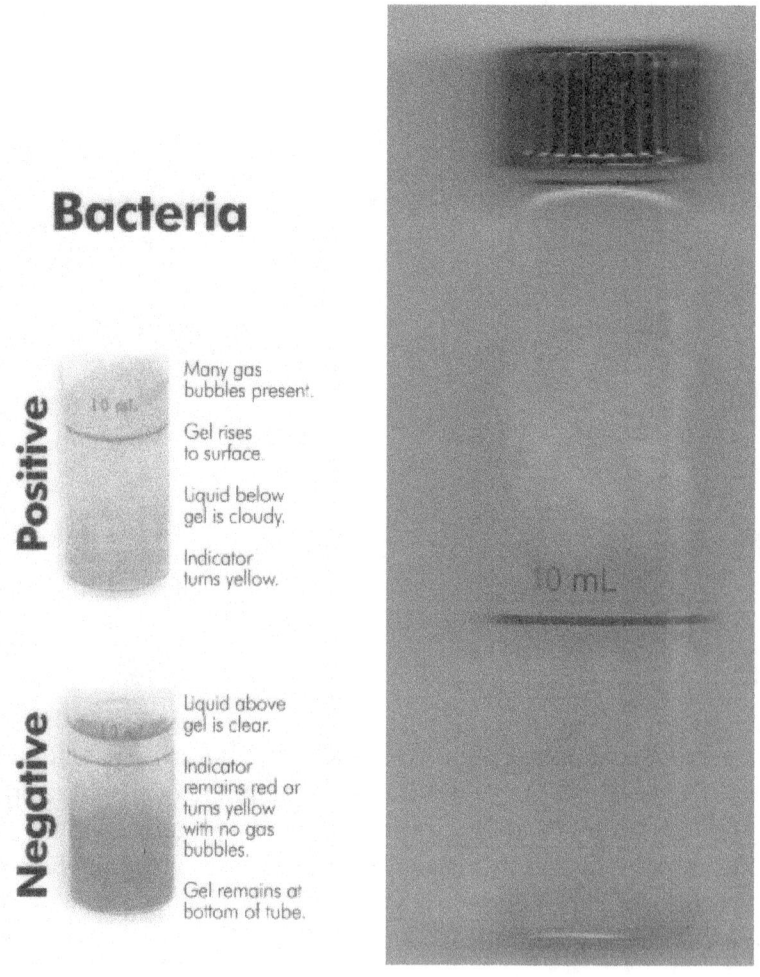

The tap water ran for 3 minutes before the water sample was taken.

Again, as the cautionary note specified, "A positive test result does not necessarily indicate fecal contamination".

Chapter 9 – Discussion of tests

The authors were very curious about the three separate tests as to the validity of the results indicating the Presence of Bacteria Coliform in the samples of tap water taken, where they all appeared to have: "many gas bubbles present", "Gel rises to surface", "Liquid below gel is cloudy" and "Indicator turns yellow" using the La Motte Urban Water Quality Test Kit 5918.

As such, the authors wrote to the tech department at La Motte seeking an explanation [see **Appendix 6a**, cited below].

La Motte's "tech" department replied with their explanation [see **Appendix 6b**, cited below].

As they mentioned:

In order to maintain sterility and avoid contamination, the Total Coliform test vial and tablet should not be opened or handled prior to use. The tablet in the bottom of the vial should be white; it will turn brown or gray if it gets contaminated. It is also important to collect your sample water in a clean, sterile container.

Care should be taken to avoid cross contamination of the sample when collecting, handling and filling the test vial [see **Appendix 6b**, cited below].

The authors felt that "cross-contamination" was not an issue or explanation, as three separate samples were taken [07 August, 09 August and 06 September, 2011 following La Motte's Bacteria Coliform procedures "to the letter" [see **Figure 4**, appearing below].

Figure 4: Urban Water Quality Test Kit 5918

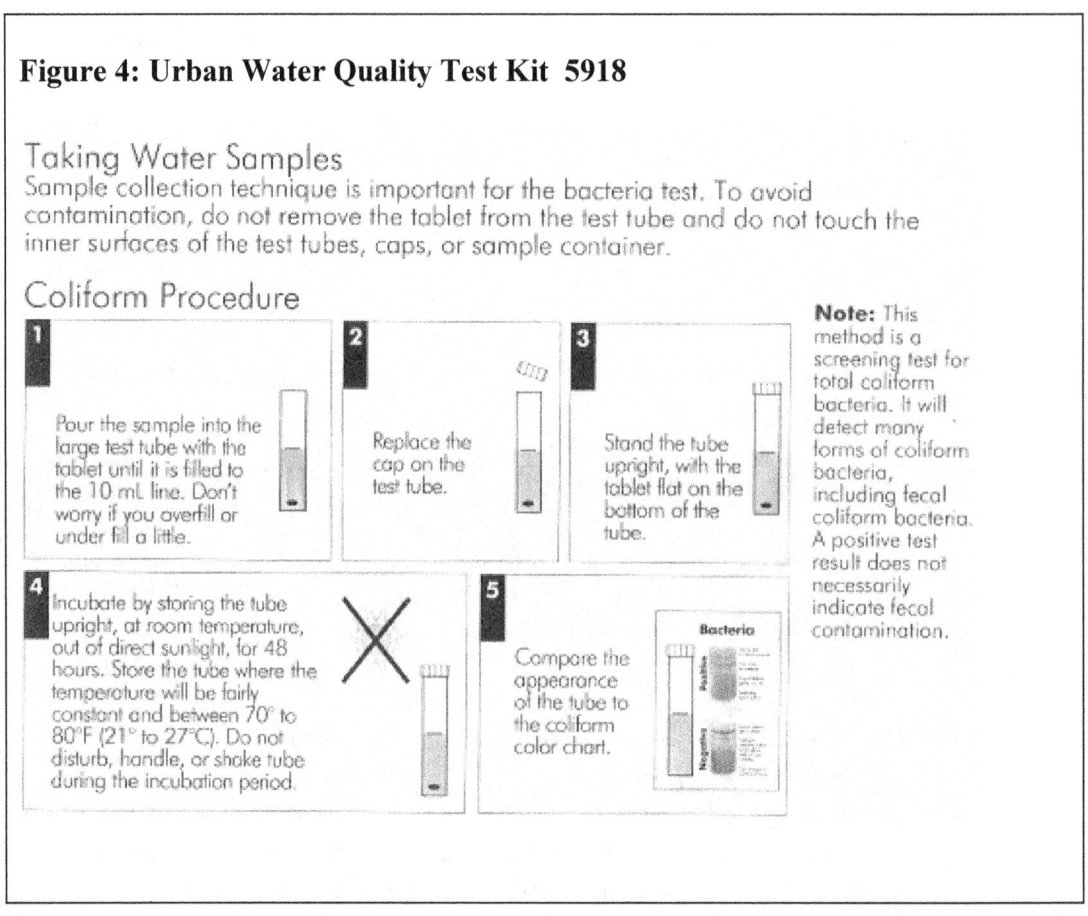

In addition, the authors had a local professional laboratory test their tap water sample on 11 August, 2011.

Total Coliform and E. Coli results appear in **Figure 8**, appearing below.

Figure 8: Local Lab Test Results:

DRINKING WATER from tap water SAMPLE TAKEN 11 August, 2011

Product/Matrix: TC-MCOLIMF

Desc: Total Coliform mcoli blue MF

Parameter

E. Coli

Final Result	Raw Value
<1	0
Units	Analyzed
CFU/100m	11-AUG-11

Parameter

Total Coliforms

Final Result	Raw Value
<1	0
Units	Analyzed
CFU/100m	11-AUG-11

Total Coliform, EColi Mcoli Blue & HPC			
Escherichia Coli mcoli blue MF			
E. Coli	<1	CFU/100mL	10-AUG-11
Heterotrophic Plate Count			
Heterotrophic Plate Count	<10	CFU/mL	10-AUG-11
Total Coliform mcoli blue MF			
Total Coliforms	<1	CFU/100mL	10-AUG-11

Appendix 6a

LaMotte Company
802 Washington Ave
Chestertown, MD 21620 USA
800-344-3100
410-778-6394 Fax
sbyerly@lamotte.com
tech@lamotte.com

Hello,

We recently moved to our new location and wanted to test the tap water.

We bought one of your Urban Water Quality Test Kit 5918 .

To our surprise, the coliform bacteria results were apparently "through the roof" [see attached]?

However, when we had a local lab check the results, they said there was zero "0" coliform bacteria [see attached]?

I'm wondering if your Urban Kit is <u>more sensitive</u> or is it measuring something else [a different coliform]?

Can you explain?

Thanks

Jeremy Mallenby, BA, BSc

Appendix 6b

RE: question - Urban Water Quality Test Kit
From: Charlie Gloyd (cgloyd@lamotte.com)
Sent: August-17-11 8:00:23 AM
To: thegoodones@live.ca (thegoodones@live.ca)

Dear Sir,

Thank you for selecting LaMotte analytical products.

The Total Coliform test used in the Urban Water Quality kit is intended to screen for the presence of Total Coliform bacterias.

Following are a few guidelines for completing the Total Coliform test.

In order to maintain sterility and avoid contamination, the Total Coliform test vial and tablet should not be opened or handled prior to use. The tablet in the bottom of the vial should be white; it will turn brown or gray if it gets contaminated. It is also important to collect your sample water in a clean, sterile container.

Care should be taken to avoid cross contamination of the sample when collecting, handling and filling the test vial.

After filling the Total Coliform vial and capping tightly, set it upright in a location away from direct sunlight with temperatures between 21 & 27 C. DO NOT shake the sample and check after "incubating" for 48 hours. Any growth that occurs after the 48 hour incubation period should be ignored.

Since this kit is an absence/presence test, we generally recommend filling more one test vial with a water sample.

In addition to this general Total Coliform screening test, we also offer ColiQuant kits that will test BOTH Total Coliform bacteria and E. coli bacteria. These kits are more sensitive and will provide colony counts for each type of bacteria.

Below is a link to our website :
http://lamotte.com/microbiological/product_line/coliquant_ez_mf.html

I trust this information is helpful,

Charlie Gloyd
Lamotte Technical Service
Phone: 800 344 – 3100 x 7028
E-mail: cgloyd@lamotte.com

Chapter 10 – Suggestion by La Motte

One of the suggestions made by La Motte's "tech" department was to use their ColiQuant kits that will test BOTH Total Coliform bacteria and E. coli bacteria and "these kits are more sensitive and will provide colony counts for each type of bacteria" [see **Appendix 6b**, cited above].

The authors will be re-doing their experiment(s) on Winnipeg, Manitoba tap water when they receive these new kits from La Motte.

In the mean-time, the authors were interested in any possible sources of Coliform Bacteria as described above.

As cited, "Coliform bacteria cause no observable odor, taste, or color change in water. Testing is the only way to determine if they are present."[1]

"Coliform bacteria are a commonly used bacterial indicator of sanitary quality of foods and water. They are defined as rod-shaped Gram-negative non-spore forming bacteria which can ferment lactose with the production of acid and gas when incubated at 35-37°C"[2] [see **Appendix 7**, cited below].

"Escherichia coli (E. coli), a rod-shaped member of the coliform group, can be distinguished from most other coliforms by its ability to ferment lactose at 44°C in the fecal coliform test, and by its growth and color reaction on certain types of culture media."[3]

"Coliform bacteria can result from the environment naturally or be fecal in origin."[4]

"Not all coliform bacteria are harmful."[5]

"However, if found in water, their presence suggests other disease causing organisms may exist in your drinking water supply."[6]

"It is common to have contamination when maintenance has been performed on the water supply system."[7]

"Faulty in-ground sewage disposal systems may also be the source of contamination."[8]

There is confusion, as noted, "Coliforms are a sub-group of bacteria that are often mentioned in reference to water quality. There are numerous bacteria that fall under the coliform heading, which leads to some confusion when discussing bacteria types and numbers."[9]

"A total coliform test measures just that, all coliform bacteria. There are numerous coliform species which occur naturally in the environment. Most are considered harmless, however some of the bacteria are harmful and indicate fecal contamination."[10]

"Total coliform bacteria testing is used to screen water quality because the bacteria are easy to culture in a lab setting. The non-fecal types of bacteria tend to survive longer in the environment than the those associated with fecal material. This means the window of opportunity for finding coliform bacteria in a water sample is longer than if just fecal coliform were being tested."[11]

"One group of coliform bacteria, fecal coliforms, is associated with human and animal waste. E.coli is a type of fecal coliform bacteria found in the intestines of both humans and animals."[12]

"The presence of E.coli in water is an indication of recent sewage or animal waste contamination. Water contaminated with E.coli is unsuitable for drinking."[13]

Could such be a source of such apparent contamination in the La Motte's Bacteria Coliform tests cited above.

Under Manitoba's Manitoba Water Stewardship, "Section 9 of The Drinking Water Safety Act requires periodic, third-party assessments of public and semi-public water systems. The assessment considers the condition and adequacy of the water system's infrastructure (supply, treatment, storage, distribution), and the quality and vulnerability of the water source and treated water."[14]

With respect to the city of Winnipeg, the steps they use to ensure "drinking water is high-quality and safe"[15]:

•"We work with the First Nation communities in the Shoal Lake area, the federal government and the provincial governments of Manitoba and Ontario to make sure that Shoal Lake, our source of water, is protected.

•We routinely test our water each step of the way, from Shoal Lake to the tap.

•We treat the water at our new state-of-the-art drinking water treatment plant.

•We disinfect the water with chlorine to kill harmful viruses and bacteria, such as E. coli.

•We use ultraviolet light to protect against waterborne parasites, such as Cryptosporidium (crip-toe-spor-ID-ee-um) and Giardia (GEE-ar-dee-ah).

•We add orthophosphate to the water to reduce the amount of lead getting into tap water from lead pipes.

•We test for chlorine at the three pumping stations in the city, 24 hours a day, 365 days a year. We test the chlorine levels at other places, such as water mains, every week.

•We take weekly samples for bacteria at over 60 places throughout the system and test them according to provincial regulations. Each year, we test more than 3,100 water samples for bacteria – this is 70% more than required. Test results for our water have always been within the acceptable range for bacteria. A nationally accredited laboratory conducts the bacteria tests.

•We test at least monthly for the microscopic parasites, Cryptosporidium and Giardia, even though there is no requirement for this type of testing. These parasites are found in most rivers and lakes. A nationally accredited laboratory recognized internationally as expert in the study of parasites tests our water samples for Cryptosporidium and Giardia."[16]

The program sounds very impressive.

The question is, what would raw sewage do to water quality in Winnipeg, Manitoba, Canada?

Appendix 7

Dan Obrecht, Coliform bacteria
http://www.lmvp.org/Waterline/winter2003/coliform.htm

The major groups of bacteria are defined by shape. Coliform bacteria
fall into the bacillus, or rod-shaped group.

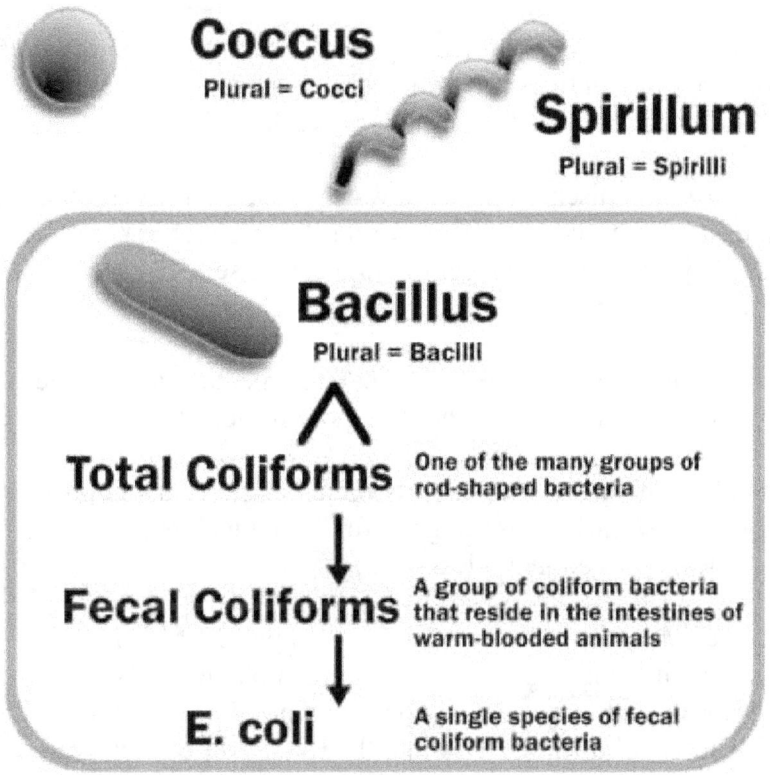

Chapter 11 – Raw sewage spill

The question is, what would raw sewage do to water quality in Winnipeg, Manitoba, Canada [see **Appendix 8a to 8d**, cited below].

For example, as cited by the CBC News [see **Appendix 8a**, cited below], "More than one billion litres of partially treated sewage have flowed into the Red River over the past few weeks, due to a major glitch at a Winnipeg waste treatment plant, according to city officials."[17]

"One of the four sewage treatment processes at the South End Water Pollution Control Plant has not been working since Oct. 7, and city staff are stumped as to how and why it broke down."[18]

"Right now we are discharging wastewater to the Red River from the effluent of the plant that is in excess of our license requirement," Mike Shkolny, the city's manager of engineering services, told reporters.[19]

"What is malfunctioning is the biological treatment stage, in which microorganisms eat organic material in the sewage."[20]

"For an unknown reason, these microorganisms suddenly stopped thriving on Oct. 7, 2011, upsetting the full biological treatment process," the city stated in a release.[21]

As further cited by the Winnipeg Sun [see **Appendix 8b**, cited below], "One of Winnipeg's sewage treatment plants has been spewing wastewater that contains up to 10 times the recommended amount of fecal coliform bacteria into the Red River for nearly a month."[22]

"The problem, which one city official called "a major upset," was first detected Oct. 7, and since then has allowed between 50 million and 60 million litres of only partly treated wastewater to flow into the river every day. Officials are baffled as to the cause, and say it could be at least a month before the problem is completely fixed."[23]

"This would be a major upset and we're having trouble bringing the plant back to normal operation," said Kelly Kjartanson, manager of environmental standards for the city.[24]

"To the city's surprise, all the bacteria in the tank have died, meaning there is nothing alive in the tank to eat the harmful organics."[25]

This apparently wasn't the first time that raw sewage has been spilled. As cited in June in the Winnipeg Sun also reported such a spill [see **Appendix 8c**, cited below], "over two-million litres of untreated sewage was discharged into the Assiniboine River last month over a five-day period."[26]

"It was the largest sewage overflow in Winnipeg since the city's massive sewage spill in 2002."[27]

"That's on top of 17 smaller spills that occurred this year between March and April."[28]

"It's all due to Winnipeg's outdated combined sewer system, which diverts raw sewage into our rivers every time it rains, during spring runoff and when pipes get clogged, like they did last month."[29]

"Despite that, there was nothing in (Provincial Premier) Selinger government's master plan released last week to "save Lake Winnipeg" that deals with the city's combined sewer problem."[30]

"In fact, it doesn't even mention it."[31]

Incredibly, there are apparently other such episodes in Winnipeg, Manitoba, Canada.

As cited [see **Appendix 8d**, cited below], "SWOLLEN river levels coupled with some moderate precipitation have led to a high possibility of some raw sewage discharge into the Red and Assiniboine Rivers over the last 48 hours."[32]

"Winnipeg's aging combined sewer system is such that when water levels in the rivers are high, the capacity of the system is low. When the rains come, the extra runoff adds more stress on the system and puts unprotected basements at risk."[33]

"The city averages about 22 overflows annually, which amounts to about 1.14 million cubic metres of raw sewage spilled into the rivers."[34]

In fact the situation seems to be so dire that others have reported concerns.

For example, one article refers to the "dirty municipal infrastructure" [see **Appendix 9**, cited below], "as the United States becomes a nation of 300 million, the country's older cities face the reality of overpopulation, crumbling infrastructures, and the health concerns raised by both, especially those related to the availability of fresh water."[35]

"Investigations conducted in the last five years suggest that a substantial proportion of waterborne disease outbreaks, both microbial and chemical, are attributable to problems within distribution systems," said the National Research Council in a report released in December for the Environmental Protection Agency.[36]

Footnotes

1 - 2. *Water Test, Inc. tests for the following contaminants*
http://www.watertestinc.com/contaminants.html

3. *Coliform bacteria*
http://en.wikipedia.org/wiki/Coliform_bacteria

Also see: American Public Health Association (APHA), *Standard Methods for the Examination of Water and Wastewater (19th ed.)*, APHA, Washington, DC (1995).

4 - 8. *Coliform Bacteria*
http://docs.google.com/viewer?a=v&q=cache:oZUrLLbaTmoJ:www.saskh2o.ca/PDF-WaterCommittee/ColiformBacteria.pdf+coliform+bacteria&hl=en&gl=ca&pid=bl&srcid=ADGEESgOalMxtLv8G6QZhSL3mgidGqH_F3FlhnsGeNOEpZLugc5honP_6NdQD_rAbkqaT0n03fTR_HyEVIhwkp42htSeGEAkvbmOAXoObgjtBwD1bSBvpmSJnZ3-ibpdNrzzHaqX8Hi8&sig=AHIEtbQUFW5zGQj7QhvhDk6412spbiY0LA

9 - 11. Dan Obrecht, Coliform bacteria
http://www.lmvp.org/Waterline/winter2003/coliform.htm

12 - 13. *Water Test, Inc. tests for the following contaminants*
http://www.watertestinc.com/contaminants.html

14. *Manitoba Water Stewardship*
Regulatory Information - Approval Requirements
http://www.gov.mb.ca/waterstewardship/odw/reg-info/approvals/index.html

15 - 16. *Winnipeg's drinking water quality*
http://www.winnipeg.ca/waterandwaste/water/quality.stm

17 - 21. *Partially treated sewage flows from Winnipeg plant*
CBC News

Posted: Nov 2, 2011 4:36 PM CT
http://www.cbc.ca/news/canada/manitoba/story/2011/11/02/winnipe
g-wastewater-plant-problem.html

22 - 25. *Sewage treatment plant spewing into Red*
By Paul Turenne,Winnipeg Sun
First posted: Wednesday, November 02, 2011
http://www.winnipegsun.com/2011/11/02/sewage-treatment-plant-
spewing-into-red

26 - 31. *Make sewers election issue*
By Tom Brodbeck, Winnipeg Sun
First posted: Monday, June 06, 2011 10:46 PM CDT
http://www.winnipegsun.com/2011/06/06/make-sewers-election-
issue

32 - 34. *Raw sewage may hit rivers, city warns*
By: Staff Writer
Winnipeg Free Press - PRINT EDITION
Posted: 04/4/2010 1:00 AM | Comments: 0 (including replies) | Last
Modified: 04/4/2010 12:00 PM | Updates
http://www.winnipegfreepress.com/local/raw-sewage-may-hit-rivers-
city-warns-89863817.html

35 - 36. *Most tap water polluted by dirty municipal infrastructure*
By Beau Hodai , Thursday, January 25, 2007
http://www.naturalnews.com/021504.html

Appendix 8a

Partially treated sewage

Partially treated sewage flows from Winnipeg plant
CBC News
Posted: Nov 2, 2011 4:36 PM CT
http://www.cbc.ca/news/canada/manitoba/story/2011/11/02/winnipeg-wastewater-plant-problem.html

More than one billion litres of partially treated sewage have flowed into the Red River over the past few weeks, due to a major glitch at a Winnipeg waste treatment plant, according to city officials.

One of the four sewage treatment processes at the South End Water Pollution Control Plant has not been working since Oct. 7, and city staff are stumped as to how and why it broke down.

As a result of the malfunction, effluent coming from the plant — which flows into the Red River — is currently being treated to just 50 per cent of how it would normally be treated, officials said Wednesday.

"Right now we are discharging wastewater to the Red River from the effluent of the plant that is in excess of our license requirement," Mike Shkolny, the city's manager of engineering services, told reporters.

Shkolny said it may sound like a lot of sewage is flowing into the Red River, but he stressed that it's actually a relatively small amount.

He added that it's not raw sewage going into the river, but partially treated sewage.

Biological stage not working

What is malfunctioning is the biological treatment stage, in which microorganisms eat organic material in the sewage.

"For an unknown reason, these microorganisms suddenly stopped thriving on Oct. 7, 2011, upsetting the full biological treatment process," the city stated in a release.

The first two treatment stages — in which grit, sediment and grease are removed from the waste — are working normally.

However, the final ultraviolet disinfection stage is operating at "reduced effort" because the biological treatment stage isn't working, officials say.

Shkolny said city engineers have been unsuccessful in trying to fix the problem so far.

A team of experts has been assembled to work on the issue, but it could be at least another month before they figure out why the treatment process broke down.

Does not meet provincial standards

City officials say this is the first time such a disruption has happened in the history of Winnipeg's sewage treatment plants.

The South End plant, which opened in 1974, treats 60 million litres of sewage a day.

Considering the sewage has been partially treated for the past 26 days, that means more than one billion litres have gone into the river to date.

Shkolny said the problem is not a public health issue, but provincial regulators have been notified because the partially treated effluent does not meet Environment Act licence requirements for the plant.

Fewer people are going in and around the Red River at this time of year, but those who come into contact with the river water should wash their hands thoroughly afterward, he said.

"Full-body contact immersion would not be recommended," Shkolny said.

"However, at this point in time, there's not much recreation going on in the river, so the risk to public health relative to swimming or boating or water-skiing is quite small."

Fishing outfitter surprised

Anyone who catches fish from the river should wash and boil their catch before eating it, he added.

Stu McKay, who owns a fishing outfitting store in nearby Lockport, Man., said the plant malfunction is an example of people's disrespect for water.

McKay said he was surprised to learn that partially treated sewage has been going into the Red River, which runs near his store, Cats on the Red.

"It's amazing, actually, in this day and age that we don't have systems put in place to prevent these types of things [from] happening. I mean, what does it take?" McKay told CBC News.

"Do we not respect water, or should we not be giving it more respect than what we have been in this day and age, knowing that it's the most important resource we have on the planet?"
Make sewers election issue 14

Appendix 8b

Partially treated sewage

Sewage treatment plant spewing into Red
By Paul Turenne, Winnipeg Sun
First posted: Wednesday, November 02, 2011
http://www.winnipegsun.com/2011/11/02/sewage-treatment-plant-spewing-into-red

One of Winnipeg's sewage treatment plants has been spewing wastewater that contains up to 10 times the recommended amount of fecal coliform bacteria into the Red River for nearly a month.

The problem, which one city official called "a major upset," was first detected Oct. 7, and since then has allowed between 50 million and 60 million litres of only partly treated wastewater to flow into the river every day. Officials are baffled as to the cause, and say it could be at least a month before the problem is completely fixed.

"This would be a major upset and we're having trouble bringing the plant back to normal operation," said Kelly Kjartanson, manager of environmental standards for the city.

"If you come into contact with the river, wash your hands," said Mike Shkolny, manager of the city's engineering division.

The plant in question is the south end sewage treatment plant, located near St. Mary's Road and the south Perimeter Highway. It handles 20% to 25% of Winnipeg's sewage.

The problem relates to the third of four steps wastewater goes through at the plant. After having grit, sediments and other large particles removed, sewage is sent to a tank filled with what Shkolny called "a bacterial cocktail" that sees beneficial bacteria consume the harmful organics contained in the sewage.

To the city's surprise, all the bacteria in the tank have died, meaning there is nothing alive in the tank to eat the harmful organics, so they're being discharged into the river.

The water now flowing into the Red contains about double the ammonia it did prior to the malfunction and fecal coliform counts — most commonly the E. coli bacteria — that are not only well above recreational guidelines, but amounts in violation of the city's environmental licence.

The city is licensed to keep the count below 200 bacteria per 100 ml of water, and current levels are between 1,000 and 2,000, city officials said.

The plant's managers still aren't sure what went wrong but have assembled a team of experts to try to solve the problem.

Shkolny said the city is only making the problem public now because it expected to be able to solve the issue earlier but has since realized its efforts were unsuccessful.

"Our hope would be to be see some improved quality soon and that in a month, maybe longer, we'll be back to normal," he said.

Appendix 8c

Partially treated sewage

Make sewers election issue
By Tom Brodbeck, Winnipeg Sun
 First posted: Monday, June 06, 2011 10:46 PM CDT
http://www.winnipegsun.com/2011/06/06/make-sewers-election-issue

Over two-million litres of untreated sewage was discharged into the Assiniboine River last month over a five-day period.

It was the largest sewage overflow in Winnipeg since the city's massive sewage spill in 2002.

That's on top of 17 smaller spills that occurred this year between March and April.

It's all due to Winnipeg's outdated combined sewer system, which diverts raw sewage into our rivers every time it rains, during spring runoff and when pipes get clogged, like they did last month.

Despite that, there was nothing in the Selinger government's master plan released last week to "save Lake Winnipeg" that deals with the city's combined sewer problem.

In fact, it doesn't even mention it.

I don't get that.

On May 20, the city's 311 service got an e-mail at 2:11 p.m. that reported a raw sewage discharge into Sturgeon Creek near Lonsdale Drive just west of Grace General Hospital.

Unfortunately, the 311 system broke down and raw sewage poured into the creek and river for nearly five days without the city responding to it. You might want to look into this one, Sam.

The wastewater collection branch wasn't notified until Wednesday May 25 — following a long-weekend — and a crew was eventually dispatched that day.

They found a blockage of grease and rags in the sewer that caused raw sewage to build up and overflow into the creek.

It was a major screw-up — and a lot crap that went into our rivers and lakes.

"Lag time between notification and resolution due to oversight in internal protocol," the city's incident report says. "Response process reviewed and will be improved for future similar events."

Let's hope so.

But better than that, why doesn't the provincial government take the lead on this and sit down with the city to hammer out a funding deal that would fix this problem over time?

Isn't that the real solution here?

I didn't see any of that in the NDP's "Save Lake Winnipeg" plan. And when asked about why it wasn't included in the plan, Premier Greg Selinger said the combined sewer problem isn't as high of a priority for him as improving Winnipeg's wastewater treatment facilities.

Why not fix both? Or at least make a commitment to work with the city and Ottawa to devise a 15-20 year plan to solve the problem.

It's called leadership.

How many times are we going to allow major spills like last month's sewage discharge and the 30 or 40 other sewage dumps that occur every year?

Does anyone really believe we can save Lake Winnipeg by ignoring the combined sewer problem?

I don't know how any government can claim to be "green" when it says the combined sewer system problem isn't a priority.

Manitoba's Clean Environment Commission recommended in 2003 that the combined sewer problem be addressed immediately, including better monitoring of spills and swift action to fix the areas in greatest need.

Eight years later, almost nothing has been done.

The problem is that fixing combined sewers is an expensive proposition — between $500 million and $1.5 billion depending on how many spills the city wants to eliminate.

Which is why senior levels of government — led by the province — have to get involved. They have to be part of the funding solution. The city can't do it on its own.

Fixing combined sewers isn't the only thing that needs to be done to save Lake Winnipeg. It's one of many.

But it's shocking it wouldn't be part of any provincial government plan to save one of Manitoba's most precious natural assets.

Hopefully it will become a campaign issue during this fall's provincial election

Appendix 8d

Partially treated sewage

Raw sewage may hit rivers, city warns
By: Staff Writer
Winnipeg Free Press - PRINT EDITION
Posted: 04/4/2010 1:00 AM | Comments: 0 (including replies) | Last
Modified: 04/4/2010 12:00 PM | Updates
http://www.winnipegfreepress.com/local/raw-sewage-may-hit-rivers-
city-warns-89863817.html

*SWOLLEN river levels coupled with some moderate precipitation have
led to a high possibility of some raw sewage discharge into the Red
and Assiniboine Rivers over the last 48 hours.*

*Winnipeg's aging combined sewer system is such that when water
levels in the rivers are high, the capacity of the system is low. When the
rains come, the extra runoff adds more stress on the system and puts
unprotected basements at risk.*

*The city says most of the combined rain and waste water makes it to
the treatment facility. The untreated solution that doesn't, though, finds
its way into the river through various overflow points along the two
rivers.*

*The city averages about 22 overflows annually, which amounts to
about 1.14 million cubic metres of raw sewage spilled into the rivers.*

Appendix 9

Dirty municipal infrastructure

Most tap water polluted by dirty municipal infrastructure
Thursday, January 25, 2007 by: Beau Hodai
http://www.naturalnews.com/021504.html

(NaturalNews) As the United States becomes a nation of 300 million, the country's older cities face the reality of overpopulation, crumbling infrastructures, and the health concerns raised by both, especially those related to the availability of fresh water.
Eric Goldstein, a spokesman for the Natural Resources Defense Council, has stated that the water distribution systems of cities such as Chicago, Denver, Philadelphia and New York are in urgent need of repair.

The antiquated water delivery systems in these cities are comprised of nearly 1 million miles of piping, mostly made of iron. As the iron pipes corrode, clean water flowing through them becomes contaminated with rust. Over time the pipes also rupture, causing not only water loss, but the introduction of pollutants and diseases from the ground.

"Investigations conducted in the last five years suggest that a substantial proportion of waterborne disease outbreaks, both microbial and chemical, are attributable to problems within distribution systems," said the National Research Council in a report released in December for the Environmental Protection Agency.

There are 170,000 public water distribution systems at work nationwide, and municipalities spend more than $50 million each year to supply clean drinking water in accordance with the Safe Drinking Water Act of 1974.

"If you clean up water and then put it into a dirty pipe, there's not much point," said Montana State University microbiologist and water research scientist, Timothy Ford. "I consider the distribution system to be the highest risk and the greatest problem we are going to be facing in the future," said Ford.

Jack Hossbuhr, executive director of the American Water Works Association, estimates that the cost of replacing existing pipelines over the next 20 to 30 years is going to cost water utility companies some $250 to $350 billion

Chapter 12 – Water quality before sewage spill

The question asked, what would raw sewage do to water quality in Winnipeg, Manitoba, Canada [see **Appendix 8a to 8d**, cited above].

It would be important to know what the water quality was before this raw sewage spill:

"More than one billion litres of partially treated sewage have flowed into the Red River over the past few weeks, due to a major glitch at a Winnipeg waste treatment plant, according to city officials."[1]

"One of the four sewage treatment processes at the South End Water Pollution Control Plant has not been working since Oct. 7, and city staff are stumped as to how and why it broke down."[2]

"One of Winnipeg's sewage treatment plants has been spewing wastewater that contains up to 10 times the recommended amount of fecal coliform bacteria into the Red River for nearly a month."[3]

"The city is licensed to keep the count below 200 bacteria per 100 ml of water, and current levels are between 1,000 and 2,000, city officials said."[4]

"Over two-million litres of untreated sewage was discharged into the Assiniboine River last month over a five-day period."[5]

"It was the largest sewage overflow in Winnipeg since the city's massive sewage spill in 2002."[6]

"That's on top of 17 smaller spills that occurred this year between March and April."[7]

"It's all due to Winnipeg's outdated combined sewer system, which diverts raw sewage into our rivers every time it rains, during spring runoff and when pipes get clogged, like they did last month."[8]

The current authors ironically had had their tap water in Winnipeg, Manitoba, Canada tested prior to the raw sewage spill on 10 August, 2011.

The results of that analysis appear as **Figure 9**, appearing below.

Figure 9: Water quality before the raw sewage spill

Sample Details/Parameters	Result	Qualifier*	D.L.	Units	Extracted
Total Coliform, EColi Mcoli Blue & HPC					
Escherichia Coli mcoli blue MF					
E. Coli	<1		1	CFU/100mL	10-AUG-11
Heterotrophic Plate Count					
Heterotrophic Plate Count	<10		10	CFU/mL	10-AUG-11
Total Coliform mcoli blue MF					
Total Coliforms	<1		1	CFU/100mL	10-AUG-11
WP2 Drinking Water					
Chloride					
Chloride	23.4		0.50	mg/L	
Conductivity					
Conductivity	293		0.40	umhos/cm	
Fluoride					
Fluoride	0.70		0.10	mg/L	
Hardness - grains/Imperial gallon					
Hardness-grains /IMPgal	6.26		0.010	grn/IMPgal	
Hardness - grains/US gallon					
Hardness-grains/USgal	5.22		0.010	grn/USgal	
Hardness Calculated					
Hardness (as CaCO3)	89.2		0.30	mg/L	
Nitrate as N					
Nitrate-N	<0.050		0.050	mg/L	
Nitrate+Nitrite					
Nitrate and Nitrite as N	<0.071		0.071	mg/L	
Nitrite as N					
Nitrite-N	<0.050		0.050	mg/L	
Sulfate					
Sulfate	55.1		0.50	mg/L	
TDS (Calculated from EC)					
TDS (Calculated from EC)	190		20	mg/L	
Total Metals by ICP-MS					
Arsenic (As)-Total	<0.0010		0.0010	mg/L	11-AUG-11
Barium (Ba)-Total	0.0151		0.00050	mg/L	11-AUG-11
Boron (B)-Total	<0.030		0.030	mg/L	11-AUG-11
Calcium (Ca)-Total	24.8		0.20	mg/L	11-AUG-11
Copper (Cu)-Total	0.0341		0.0020	mg/L	11-AUG-11
Iron (Fe)-Total	0.14		0.10	mg/L	11-AUG-11
Magnesium (Mg)-Total	6.63		0.050	mg/L	11-AUG-11
Manganese (Mn)-Total	0.0330		0.0010	mg/L	11-AUG-11
Potassium (K)-Total	1.48		0.10	mg/L	11-AUG-11
Sodium (Na)-Total	34.3		0.050	mg/L	11-AUG-11
Uranium (U)-Total	<0.00050		0.00050	mg/L	11-AUG-11
Zinc (Zn)-Total	<0.020		0.020	mg/L	11-AUG-11
pH					
pH	7.98		0.10	pH units	
Miscellaneous Parameters					
Ammonia as N	<0.050		0.050	mg/L	
Phosphorus (P)-Total	0.612		0.010	mg/L	

In looking at the Coliform counts, cited in **Figure 9**, it looks like the results are within limits as < 1 CFU/ml: "the maximum acceptable concentration (MAC) of Escherichia coli in public, semi-public, and private drinking water systems is none detectable per 100 mL."[9]

As cited, "the presence of excess total coliforms, as stated below, in drinking water indicates that treatment is inadequate or that the distribution system is experiencing regrowth or infiltration. Total coliforms are not necessarily an indication of the presence of faecal contamination."[10]

"Faecal coliforms in drinking water may, however, indicate the presence of faecal contamination. The presence of Escherichia coli, one species in the faecal coliform group, is a definite indicator of the presence of faeces."[11]

"Other species in the faecal coliform group (e.g., Klebsiella pneumoniae, Enterobacter cloacae) are not restricted to faeces but occur naturally on vegetation and in soils."[12]

"The MAC for coliforms in drinking water is zero organisms detectable per 100 mL. Because coliforms are not uniformly distributed in water and are subject to considerable variation in enumeration, drinking water that fulfils the following conditions is considered to be in compliance with the coliform MAC:

1. No sample should contain more than 10 total coliform organisms per 100 mL, none of which should be faecal coliforms;

2. No consecutive sample from the same site should show the presence of coliform organisms; and

3. For community drinking water distribution systems:

   a) not more than one sample from a set of samples taken from the community on a given day should show the presence of coliform organisms; and

   b) not more than 10% of the samples based on a minimum of 10 samples should show the presence of coliform organisms."[13]

The first result, "Escherichia coli mcoli blue MF", has in fact been approved by the EPA [see **Appendix 10**, cited below] [14], with:

"12.1.1 The presence of at least one blue/purple or pink/magenta colony at least 0.5 mm in diameter indicates the sample is total coliform positive."[15]

"The presence of at least one blue/purple colony indicates the sample is positive for E. coli."[16]

"The presence of at least one pink/magenta colony indicates the sample is positive for general coliforms."[17]

In looking at the Coliform counts, cited in **Figure 9**, the heterotrophic plate count is < 10 CFU/ml.

As cited, one must be careful in interpretation, "the selection of culture media, the incubation time as well as how the plates are read (ocular or under magnification) will all affect the quantitative values obtained."[18]

In this study, "since the heterotrophic plate counts were read after 2 days in the Report A and after 3 days in the Report B, these values are not fully comparable. It is however anticipated that local factors at the taps would be the major factor influencing the variability."[19]

"The information and conclusions made in the Report B is highly relevant. The results are summarized from the 4 performed sampling rounds of microbial analysis. This showed variations between the sampled cities and also variations between the sampling rounds, where also external factors, like temperature, may have played a role."[20]

Chloride is cited as 23.4 mg/L in **Figure 9**.

As cited by Chris Mechenich & Elaine Andrews, "Interpreting Drinking Water Test Results" [see **Figure 10**, appearing below], any amount over 10 mg/l is possibly due to human influences, such as road salt, fertilizers, animal or other waste.[21]

As also noted, "shallow aquifer groundwater was slightly basic on average, with moderate levels of nitrate, chloride, total dissolved solids, and conductivity. It generally contained low levels of dissolved organic carbon, sulphate, and metals. The higher levels of chloride and nitrate in shallow aquifer groundwater reflect a generally higher susceptibility of the overburden aquifer to surficial contamination by fertilizers, septic waste, and road salt."[22]

Figure 10: Chloride

Reference to: Chris Mechenich & Elaine Andrews,
        Interpreting Drinking Water Test Results
        G3558 - 4
        http://www4.uwsp.edu/cnr/gndwater/privatewells/Interp
        reting%20Drinking%20Water%20Test%20Results.pdf

Chloride

In most areas of Wisconsin, chloride in groundwater is naturally less than 10 mg/l. Some higher concentrations in limestone and sandstone aquifers in eastern Wisconsin may also be natural. Higher concentrations usually indicate contamination by septic systems, road salt, fertilizer, animal or other wastes. Chloride is not toxic, but some people can detect a salty taste at 250 mg/l. Water with high

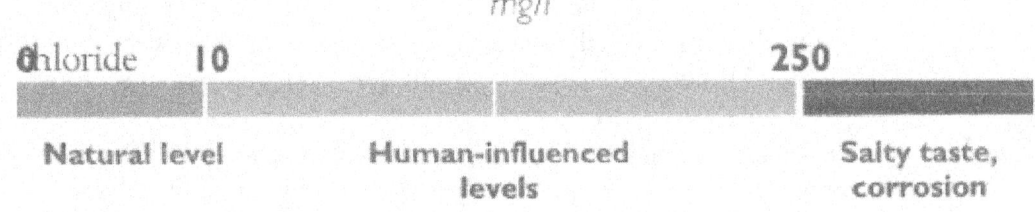

may also
have a high sodium content. High chloride may also speed up corrosion in plumbing (just as road salt does to your car).

ACCEPTABLE RESULTS: There is no health standard. Levels less than 10 mg/l are desirable. Levels more than 250 mg/l may cause a salty taste.

SOURCES: Septic systems, road salt, fertilizer, animal or other wastes.

CORRECTIVE ACTIONS: None required specifically for chloride. If elevated chloride levels are found in combination with high nitrate levels, take corrective actions indicated for nitrate.

Conductivity is cited at 293 umhos/cm in **Figure 9**.

As cited, "conductivity is a measure of the ability of water to pass an electrical current. Conductivity in water is affected by the presence of inorganic dissolved solids such as chloride, nitrate, sulfate, and phosphate anions (ions that carry a negative charge) or sodium, magnesium, calcium, iron, and aluminum cations (ions that carry a positive charge)."[23]

"Organic compounds like oil, phenol, alcohol, and sugar do not conduct electrical current very well and therefore have a low conductivity when in water."[24]

"Conductivity is also affected by temperature: the warmer the water, the higher the conductivity. For this reason, conductivity is reported as conductivity at 25 degrees Celsius (25 C)."[25]

"The basic unit of measurement of conductivity is the mho or siemens."[26]

"Conductivity is measured in micromhos per centimeter (μmhos/cm) or microsiemens per centimeter (μs/cm)."[27]

"Distilled water has a conductivity in the range of 0.5 to 3 μmhos/cm."[28]

"The conductivity of rivers in the United States generally ranges from 50 to 1500 μmhos/cm."[29]

"Studies of inland fresh waters indicate that streams supporting good mixed fisheries have a range between 150 and 500 μhos/cm."[30]

"Conductivity outside this range could indicate that the water is not suitable for certain species of fish or macroinvertebrates."[31]

"Industrial waters can range as high as 10,000 μmhos/cm."[32]

In addition, "In comparison, bedrock aquifer groundwater was slightly higher levels of conductivity, total dissolved solids, and sulphate were most likely a natural characteristic of the bedrock aquifer."[33]

As cited by Chris Mechenich & Elaine Andrews, "Interpreting Drinking Water Test Results" [see **Figure 11**, appearing below], conductivity is about twice the hardness.[34]

Figure 11: Conductivity

Reference to: Chris Mechenich & Elaine Andrews,
Interpreting Drinking Water Test Results
G3558 - 4
http://www4.uwsp.edu/cnr/gndwater/privatewells/Interp
reting%20Drinking%20Water%20Test%20Results.pdf

Conductivity

Conductivity (specific conductance) is a measure of water's ability to conduct an electrical current. It is related to the amount of dissolved minerals in water, but it does not give an indication of which minerals are present. Conductivity (measured in μmho/cm at 25°C) is about twice the hardness (mg $CaCO_3$/l) in most

uncontaminated waters in Wisconsin. If it is much greater than two times the hardness, it may indicate the presence of contaminants such as sodium, chloride or sulfate, which may be influenced by humans or naturally-occurring. Changes in conductivity over time may indicate changing water quality.

ACCEPTABLE RESULTS: There is no health standard. A normal conductivity value is roughly twice the hardness in unsoftened water.

SOURCES: Natural and synthetic dissolved substances in the water.

CORRECTIVE ACTIONS: None specifically required for conductivity.

Hardness as CaCO3 is cited at 89.2 mg/L in **Figure 9**.

As cited, "PH and Hardness do not directly cause harm but can cause a variety of secondary effects if not treated. If water acidity is too high, corrosion can leach out lead from pipes and plumbing (see lead above), as well as damage plumbing and water heating systems. Water hardness is primarily caused by calcium and magnesium compounds.

These chemicals are not easily detected, but the negative effects include scaling of pots and pans and, if left untreated, damage to plumbing and water heaters."[35]

In addition, "water hardness is the traditional measure of the capacity of water to react with soap, hard water requiring considerably more soap to produce a lather. Hard water often produces a noticeable deposit of precipitate (e.g. insoluble metals, soaps or salts) in containers, including "bathtub ring". It is not caused by a single substance but by a variety of dissolved polyvalent metallic ions, predominantly calcium and magnesium cations, although other cations (e.g. aluminium, barium, iron, manganese, strontium and zinc) also contribute. Hardness is most commonly expressed as milligrams of calcium carbonate equivalent per litre. Water containing calcium carbonate at concentrations below 60 mg/l is generally considered as soft; 60–120 mg/l, moderately hard; 120–180 mg/l, hard; and more than 180 mg/l, very hard (McGowan, 2000). Although hardness is caused by cations, it may also be discussed in terms of carbonate (temporary) and non-carbonate (permanent) hardness."[36]

"The principal natural sources of hardness in water are dissolved polyvalent metallic ions from sedimentary rocks, seepage and runoff from soils. Calcium and magnesium, the two principal ions, are present in many sedimentary rocks, the most common being limestone and chalk. They are also common essential mineral constituents of food. As mentioned above, a minor contribution to the total hardness of water is also made by other polyvalent ions, such as aluminium, barium, iron, manganese, strontium and zinc."[37]

"A large number of studies have investigated the potential beneficial health effects of drinking-water hardness. Most of these have been ecological epidemiological studies and have reported an inverse relationship between water hardness and cardiovascular mortality. Inherent weaknesses in the ecological epidemiological study design limit the conclusions that can be drawn from these studies."[38]

"Several identified case–control and cohort studies show a negative association (i.e. protective effect) between cardiovascular mortality and drinking-water magnesium. Although this association does not necessarily demonstrate causality, it is consistent with the well-known effects of magnesium on cardiovascular function. There was no evidence of an association between total water hardness or calcium and acute myocardial infarction or deaths from cardiovascular disease (acute myocardial infarction, stroke and hypertension). There does not appear to be an association between drinking-water magnesium and acute myocardial infarction. A recent large study from the Netherlands (Leurs et al., 2010) found no overall association between calcium, magnesium or total hardness and ischaemic heart disease or stroke mortality. However, there was a reported significant inverse (beneficial) association with water magnesium for men in the highest exposure group, and the opposite effect was observed for women. Thus, further study is needed."[39]

"Exposure to hard water has been suggested to be a risk factor that could exacerbate eczema. The environment plays an important part in the etiology of atopic eczema, but specific causes are unknown. Numerous factors have been associated with eczema flare-up, including dust, nylon, shampoo, sweating, swimming and wool (Langan, 2009). A suggested explanation relative to hard water is that increased soap usage in hard water results in metal or soap salt residues on the skin (or on clothes) that are not easily rinsed off and that lead to contact irritation (Thomas & Sach, 2000). There are reports of a relationship between both 1-year and lifetime prevalence of atopic eczema and water hardness among primary-school children. Eczema prevalence trends in the secondary-school population were not significant (McNally et al., 1998). Additional studies are under way."[40]

"Point-of-entry ion exchange (water softener) devices are used in some households to remove hardness (calcium, magnesium) and iron from water. Each divalent ion (e.g. Ca^{2+} or Mg^{2+}) in the water is replaced by two sodium ions. Softening will have several aesthetically beneficial effects inside the home, such as reducing scaling in pipes, fixtures and water heaters and improving laundry and washing characteristics, but it also increases the sodium (and chloride) content of the drinking-water. Consumption of calcium and magnesium in drinking-water will, of course, be lower unless the water that is consumed is not softened or is remineralized."[41]

"Although there is some evidence from epidemiological studies for a protective effect of magnesium or hardness on cardiovascular mortality, the evidence is being debated and does not prove causality. Further studies are being conducted. There are insufficient data to suggest either minimum or maximum concentrations of minerals at this time, and so no guideline values are proposed."[42]

In addition, "Prairie provinces (mainly Saskatchewan and Manitoba) contain high quantities of calcium and magnesium, often as dolomite, which are readily soluble in the groundwater that contains high concentrations of trapped carbon dioxide from the last glaciation. In these parts of Canada, the total hardness in ppm of calcium carbonate equivalent frequently exceed 200 ppm, if groundwater is the only source of potable water. The west coast, by contrast, has unusually soft water, derived mainly from mountain lakes fed by glaciers and snowmelt. Some typical values are: Montreal 116 ppm, Calgary 165 ppm, Regina 202 ppm, Saskatoon < 140 ppm, Winnipeg 77 ppm, Toronto 121 ppm, Vancouver < 3 ppm, Charlottetown PEI 140 – 150 ppm." [43]

As cited by Chris Mechenich & Elaine Andrews, "Interpreting Drinking Water Test Results" [see **Figure 12**, appearing below], hardness is caused mostly by dissolved calcium and magnesium, the end product of dissolving limestone from soil and rock materials.

Figure 12: Hardness

Reference to: Chris Mechenich & Elaine Andrews,
  Interpreting Drinking Water Test Results
  G3558 - 4
  http://www4.uwsp.edu/cnr/gndwater/privatewells/Interp
  reting%20Drinking%20Water%20Test%20Results.pdf

Hardness

Hardness in water is caused mostly by dissolved calcium and magnesium, the end product of dissolving limestone from soil and rock materials. Hard water is beneficial to health. However, high hardness can cause lime buildup (scaling) in pipes and water

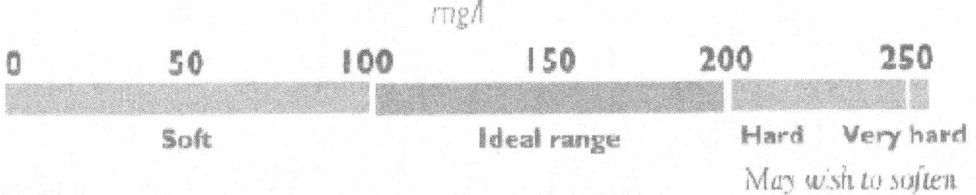

heaters. It also reacts with soap to form a "scum" which decreases soap's cleaning ability, increases bathtub ring and turns white laundry grey. Water that is naturally too soft may be corrosive. The water softening industry measures hardness in grains per gallon. One grain/gallon=17.1 mg/l $CaCO_3$

ACCEPTABLE RESULTS: Hard water is beneficial to health. However, values near 150 mg/l are ideal from an aesthetic viewpoint, if the corrosivity index is satisfactory.

SOURCES: Primarily dissolved limestone minerals from soil and rock materials.

Nitrate is cited as < 0.050 in **Figure 9**.

As cited, "under aerobic conditions, microbial activity can degrade cyanide to ammonia, which then oxidizes to nitrate. This process has been shown effective with cyanide concentrations of up to 200 parts per million. Although biological degradation also occurs under anaerobic conditions, cyanide concentrations greater than 2 parts per million are toxic to these microorganisms."[44]

Also, when animal and human wastes or field fertilizers come into contact with water, they produce nitrates and nitrites. Both are dangerous to children and can cause "Blue Baby Syndrome," a lethal form of birth defect in infants."[45]

As cited, "Nitrates and nitrites are nitrogen-oxygen chemical units which combines with various organic and inorganic compounds."[46]

"The major sources of nitrates or nitrites in drinking water include runoff from fertilizer use, sewage, and erosion of natural deposits."[47]

"Nitrogen is an important parameter to monitor."[48]

"Excessive amounts of nitrate or nitrite in water can cause methemoglobinaemia (blue baby syndrome) which can potentially be fatal."[49]

"These contaminants can also cause adult illness and produce spontaneous abortion in cows."[50]

"The EPA recommended limit for nitrates is 10 mg/L and for nitrites is 1 mg/L."[51]

As cited by Chris Mechenich & Elaine Andrews, "Interpreting Drinking Water Test Results" [see **Figure 13a & 13b**, appearing below], fertilizers, animal wastes and landfills can all contribute to elevated nitrate levels.[52]

Figure 13a: Nitrates

Reference to: Chris Mechenich & Elaine Andrews,
Interpreting Drinking Water Test Results
G3558 - 4
http://www4.uwsp.edu/cnr/gndwater/privatewells/Interp
reting%20Drinking%20Water%20Test%20Results.pdf

Nitrate

Nitrate nitrogen is a commonly used lawn and agricultural fertilizer. It is also a chemical formed in the decomposition of waste materials. If infants under six months of age drink water (or formula made with water) that contains more than 10 ppm nitrate-nitrogen, they are susceptible to methemoglobinemia, a disease which interferes with oxygen transport in the blood. High nitrate levels also suggest that other contaminants may be present. The natural level of nitrate in Wisconsin's groundwater is less than 2 mg/l. *Nitrite* is an unstable form of nitrogen which may be found in small amounts along with nitrate. Sometimes results of nitrate and nitrite are reported together.

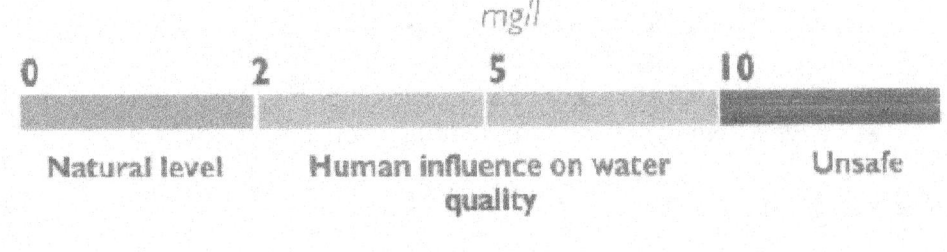

> **Figure 13b: Nitrates**
>
> Reference to: Chris Mechenich & Elaine Andrews,
> > *Interpreting Drinking Water Test Results*
> > G3558 - 4
> > *http://www4.uwsp.edu/cnr/gndwater/privatewells/Interp reting%20Drinking%20Water%20Test%20Results.pdf*
>
> ACCEPTABLE RESULTS: Labs report nitrate results either as nitrate-nitrogen or as nitrate. When reported as nitrate nitrogen (NO_3 N) or nitrate and nitrite nitrogen (NO_2 + NO_3-N) the acceptable level is less than 10 mg/l (less than 2 mg/l is preferred). When reported simply as nitrate (NO_3), the acceptable level is less than 45 mg/l.
>
> SOURCES: Fertilizer, septic system effluent, animal wastes, and landfills can all contribute to elevated nitrate levels. In most cases, elevated nitrate levels indicate general contamination of the aquifer (water-bearing formation) at that depth.

Sulfate is cited as 55.1 mg/L in **Figure 9**.

As reported, "Sulfates occur naturally in groundwater combined with calcium, magnesium and sodium as sulfate salts."[53]

"Sulfate content in excess of 250 to 500 ppm (mg/l) may give water a bitter taste and have a laxative effect on individuals not adapted to the water."[54]

"Water that smells like rotten eggs has a high level of hydrogen sulfide gas."[55]

"The gas may occur naturally in water near oil or gas fields or as the result of bacterial contamination."[56]

"Certain inorganic substances, such as sulfides, sulfites, thiosulfates, nitrites and ferrous iron are oxidized by dichromate, creating an inorganic COD, which is misleading when estimating the organic content of the wastewater"[57] [see **Appendix 11**, cited below].

"The drinking water limit for Sulfate (SO4) is 250 mg/L."[58]

However, "This regulation is not a Federally enforceable standard, but is provided as a guideline for States and public water systems. EPA estimates that about 3% of the public drinking water systems in the country may have sulfate levels of 250 mg/L or greater."[59]

"Sulfate (SO4-2) is widely distributed in natural waters, but is typically less than a few mg/L. In Northeastern Pennsylvania, the primary sources of sulfate in surface waters and groundwater include: acid mine drainage, acid deposition, and mineral oxidation. Standard set because of taste and aesthetic problems and sulfates laxative effects."[60]

"Health concerns regarding sulfate in drinking water have been raised because of reports that diarrhea may be associated with the ingestion of water containing high levels of sulfate. Of particular concern are groups within the general population that may be at greater risk from the laxative effects of sulfate when they experience an abrupt change from drinking water with low sulfate concentrations to drinking water with high sulfate concentrations."[61]

Arsenic is cited as < 0.0010 mg/L in **Figure 9**.

As cited, "Arsenic is a semi-metal element in the periodic table. It is odorless and tasteless. It enters drinking water supplies from natural deposits in the earth or from agricultural and industrial practices."[62]

"Non-cancer effects can include thickening and discoloration of the skin, stomach pain, nausea, vomiting; diarrhea; numbness in hands and feet; partial paralysis; and blindness. Arsenic has been linked to cancer of the bladder, lungs, skin, kidney, nasal passages, liver, and prostate."[63]

"EPA has set the arsenic standard for drinking water at .010 parts per million (10 parts per billion) to protect consumers served by public water systems from the effects of long-term, chronic exposure to arsenic."[64]

"According to a 1999 study by the National Academy of Sciences, arsenic in drinking water causes bladder, lung and skin cancer, and may cause kidney and liver cancer. The study also found that arsenic harms the central and peripheral nervous systems, as well as heart and blood vessels, and causes serious skin problems. It also may cause birth defects and reproductive problems."[65]

"In a February 2000 report, NRDC analyzed data compiled by the U.S. Environmental Protection Agency on arsenic in drinking water in 25 states. Our most conservative estimates based on the data indicated that more than 34 million Americans were drinking tap water supplied by systems containing average levels of arsenic that posed unacceptable cancer risks. We consider it likely that as many as 56 million people in those 25 states were drinking water with arsenic at unsafe levels -- and that's just the 25 states that reported arsenic information to the EPA."[66]

In addition, it was reported "arsenic may enter lakes, rivers or underground water naturally, when mineral deposits or rocks containing arsenic dissolve. Arsenic may also get into water through the discharge of industrial wastes and by the deposit of arsenic particles in dust, or dissolved in rain or snow."[67]

"These arsenic particles can enter the environment through:

- the burning of fossil fuels (especially coal);

- metal production (such as gold and base metal mining);

- agricultural use (in pesticides and feed additives); or

- waste burning."[68]

"Arsenic in drinking water is absorbed by the body when you swallow it, and distributed by the bloodstream. It does not enter the body through the skin or by inhalation during bathing or showering. The highest levels of arsenic are found in nails and hair, which accumulate arsenic over time. Your body gets rid of arsenic mostly through urine, with smaller amounts removed through the skin, hair, nails and sweat."[69]

"Health Canada and the International Agency for Research on Cancer consider arsenic a human cancer-causing agent. Its effects have been studied in a population in Taiwan where the drinking water contains naturally high levels of arsenic (over 0.35 ppm)."[70]

"Arsenic is one of the many chemicals for which Health Canada has set guidelines. A new guideline has been established at 0.010 milligrams per litre, and will continue to be reviewed to reflect new treatment methods and new information on health risks as they become available."[71]

"Data collected indicate that the levels of arsenic in Canadian drinking water are generally less than 0.005 milligrams per litre (0.005 parts per million - ppm), although concentrations may be higher in some areas."[72]

Barium (Ba) is cited at 0.0151 mg/L in **Figure 9**.

As cited, "Barium is a lustrous, machinable metal which exists in nature only in ores containing mixtures of elements."[73]

"It is used in making a wide variety of electronic components, in metal alloys, bleaches, dyes, fireworks, ceramics and glass."[74]

"In particular, it is used in well drilling operations where it is directly released into the ground."[75]

"In 1974, Congress passed the Safe Drinking Water Act. This law requires EPA to determine safe levels of chemicals in drinking water which do or may cause health problems. These non-enforceable levels, based solely on possible health risks and exposure, are called Maximum Contaminant Level Goals"[76] [see **Appendix 12**].

"The MCLG for barium has been set at 2 parts per million (ppm) because EPA believes this level of protection would not cause any of the potential health problems described below."[77]

"Short-term: EPA has found barium to potentially cause the following health effects when people are exposed to it at levels above the MCL for relatively short periods of time: gastrointestinal disturbances and muscular weakness."[78]

"Long-term: Barium has the potential to cause the following effects from a lifetime exposure at levels above the MCL: high blood pressure."[79]

Boron (B) is cited as < 0.030 mg/L in **Figure 9**.

As cited, "Boron is never found in the elemental form in nature. It exists as a mixture of the 10B (19.78%) and 11B (80.22%) isotopes (Budavari et al., 1989)."[80]

"Boron's chemistry is complex and resembles that of silicon (Cotton & Wilkinson, 1988)."[81]

"Elemental boron exists as a solid at room temperature, either as black monoclinic crystals or as a yellow or brown amorphous powder when impure. The amorphous and crystalline forms of boron have specific gravities of 2.37 and 2.34, respectively. Boron is a relatively inert metalloid except when in contact with strong oxidizing agents."[82]

"Sodium perborates are persalts, which are hydrolytically unstable because they contain characteristic boron–oxygen–oxygen bonds that react with water to form hydrogen peroxide and stable sodium metaborate ($NaBO_2 \cdot nH_2O$)."[83]

.

"Boric acid is a very weak acid, with a pKa of 9.15, and therefore boric acid and the sodium borates exist predominantly as undissociated boric acid [$B(OH)_3$] in dilute aqueous solution at pH <7; at pH >10, the metaborate anion $B(OH)_4^-$ becomes the main species in solution. Between these two pH values, from about 6 to 11, and at high concentration (>0.025 mol/litre), highly water soluble polyborate ions such as $B_3O_3(OH)_4^-$, $B_4O_5(OH)_4^-$, and $B_5O_6(OH)_4^-$ are formed."[84]

"The chemical and toxicological properties of borax pentahydrate $Na_2B_4O_7 \cdot 5H_2O$, borax $Na_2B_4O_7 \cdot 10H_2O$, boric acid, and other borates are expected to be similar on a molar boron equivalent basis when dissolved in water or biological fluids at the same pH and low concentration."[85]

"Boric acid and borates are used in glass manufacture (fibreglass, borosilicate glass, enamel, frit, and glaze), soaps and detergents, flame retardants, and neutron absorbers for nuclear installations. Boric acid, borates, and perborates have been used in mild antiseptics, cosmetics, pharmaceuticals (as pH buffers), boron neutron capture therapy (for cancer treatment), pesticides, and agricultural fertilizers."[86]

"Waterborne boron may be adsorbed by soils and sediments. Adsorption–desorption reactions are expected to be the only significant mechanism influencing the fate of boron in water (Rai et al., 1986). The extent of boron adsorption depends on the pH of the water and the concentration of boron in solution. The greatest adsorption is generally observed at pH 7.5 – 9.0 (Waggott, 1969; Keren & Mezuman, 1981; Keren et al., 1981)."[87]

"In natural waters, boron exists primarily as undissociated boric acid with some borate ions. As a group, the boron–oxygen compounds are sufficiently soluble in water to achieve the levels that have been observed (Sprague, 1972). Mance et al. (1988) described boron as a significant constituent of seawater, with an average boron concentration of 4.5 mg/kg."[88]

"In time–response and dose–response reproductive studies (Linder et al., 1990), adult male Sprague-Dawley rats were administered two doses in one day, with a total dose of 0 or 350 mg of boron per kg of body weight in the time–response experiment (animals were sacrificed at 2, 14, 28, or 57 days post-treatment) and a total dose of 0, 44, 87, 175, or 350 mg of boron per kg of body weight in the dose–response experiment (animals were sacrificed after 14 days). Adverse effects on spermiation, epididymal sperm morphology, and caput sperm reserves were observed during histopathological examinations of the testes and epididymis. The NOAEL for male reproductive effects in the dose–response study was 87 mg of boron per kg of body weight per day."[89]

"In a multi-generation study, doses of 0, 117, 350, or 1170 mg of boron per kg of feed (as borax or boric acid) were administered to male and female rats (Weir & Fisher, 1972). At the highest dose, rats were found to be sterile, males showed atrophied testes in which spermatozoa were absent, and females showed decreased ovulation. The NOAEL in this study was 350 mg of boron per kg of feed, quivalent to 17.5 mg of boron per kg of body weight per day."[90]

"Price et al. (1996a) did a follow-up to the Heindel et al. (1992) study in Sprague-Dawley (CD) rats to determine a NOAEL for fetal body-weight reduction and to determine whether the offspring would recover from prenatally reduced body weight during postnatal development. Boric acid was administered in the diet to CD rats on gestation days 0–20. Dams were terminated and uterine contents examined on gestation day 20. The intake of boric acid was 0, 3.3, 6.3, 9.6, 13, or 25 mg of boron per kg of body weight per day. Fetal body weights were 99, 98, 97, 94, and 88% of controls for the low- to high-dose groups, respectively. Incidences of short rib XIII (a malformation) or wavy rib (a variation) were increased in the 13 and 25 mg of boron per kg of body weight per day dose groups relative to control litters. There was a decreased incidence of rudimentary extra rib on lumbar 1 (a variation) in the high-dose group that was deemed biologically but not statistically significant. The NOAEL in this study was 9.6 mg of boron per kg of body weight per day, based on a decrease in fetal body weight at the next higher dose."[91]

"Developmental toxicity and teratogenicity of boric acid in rabbits were investigated by Price et al. (1996b) at doses of 0, 11, 22, or 44 mg of boron per kg of body weight per day, given by gavage. Frank developmental effects in rabbits exposed to 44 mg of boron per kg of body weight per day included a high rate of prenatal mortality, an increased number of pregnant females with no live fetuses, and fewer live fetuses per live litter on day 30. At the high dose, malformed live fetuses per litter increased significantly, primarily because of the incidence of fetuses with cardiovascular defects, the most prevalent of which was interventricular septal defect. Skeletal variations observed were extra rib on lumbar 1 and misaligned sternebra. The NOAEL for maternal and developmental effects was 22 mg of boron per kg of body weight per day."[92]

"Concentrations of boron found in drinking-water from Chile, Germany, the United Kingdom, and the USA ranged from 0.01 to 15.0 mg/litre, with most values clearly below 0.4 mg/litre"[93] [see **Appendix 13**].

"These values are consistent with ranges and means observed for groundwater and surface waters. This consistency is supported by two factors: (i) boron oncentrations in water are largely dependent on the leaching of boron from the surrounding geology and wastewater discharges, and (ii) boron is not removed by conventional drinking-water treatment methods."[94]

As cited, the recommended guidelines for Boron[95]:

"Drinking Water   5.0 mg/L

Aquatic Life - Freshwater   1.2 mg/L
Aquatic Life - Marine   1.2 mg/L

Wildlife   5.0 mg/L

Livestock Watering   5.0 mg/L"[96]

Calcium is cited at 24.8 mg/L in **Figure 9**.

"Sulfates occur naturally in groundwater combined with calcium, magnesium and sodium as sulfate salts."[97]

"Water hardness is primarily caused by calcium and magnesium compounds. These chemicals are not easily detected, but the negative effects include scaling of pots and pans and, if left untreated, damage to plumbing and water heaters."[98]

"Our drinking water is the best medium bar none for getting the correct absorption of calcium … calcium is an essential component in the life-long process of laying down new bone structure. Later in life, calcium helps our bodies maintain bone mass. Before you reach thirty years of age, more bone is made than lost; after thirty, this trend reverses and a healthy diet truly is vital to ensure that your body stays strong and vital through your later years."[99]

"The calcium concentration of water varies from 1 to 135 mg/L across the USA and Canada. Most spring waters were found to have a relatively low calcium concentration, with an average of 21.8 mg/L. Purified waters contain a negligible calcium concentration. Mineral waters, on the other hand, were generally found to contain higher calcium concentrations, an average of 208 mg/L of calcium. Filtration was found to remove a considerable amount of calcium from the water, removing 89% on average."[100]

"Calcium concentration in water varied substantially from different sources in the USA and Canada. Bottled waters presented with concentrations of calcium covering a very large range. Certain tap and bottled waters present with concentrations of calcium sufficient to exhibit a deleterious effect on bisphosphonate treatment. Alternatively, certain waters may be used as a source of calcium that may provide over 40% of the recommended daily intake for calcium."[101]

Copper (Cu) is cited at 0.0341 mg/L in **Figure 9**.

As cited, "natural water usually contains very little lead. Contamination generally occurs in the water distribution system or in the pipes of a home or facility. Lead pipes, brass faucets and lead solder used to join copper pipes are the culprits. If your home was built before 1986 when the nationwide ban on lead pipes and lead solder went into effect, it is likely to have lead-soldered plumbing."[102]

"Copper is a metal found in natural deposits such as ores containing other elements."[103]

"Copper is widely used in household plumbing materials."[104]

"In 1974, Congress passed the Safe Drinking Water Act. This law requires EPA to determine the level of contaminants in drinking water at which no adverse health effects are likely to occur. These non-enforceable health goals, based solely on possible health risks and exposure over a lifetime with an adequate margin of safety, are called maximum contaminant level goals (MCLG)."[105]

"The MCLG for copper is 1.3 mg/L or 1.3 ppm."[106]

As further cited, "copper occurs free and combined in nature in many minerals. Copper may exist in natural waters and effluents as a soluble salt or as suspended solids."[107]

"The common sources of copper in drinking water are from corrosion of household plumbing systems or erosion of natural deposits."[108]

"A small amount of copper is essential for plants and animals."[109]

"Concentrations exceeding 0.1 mg/L are also useful for controlling algae and plankton growth."[110]

"Quantities ranging from 0.02 - 0.1 mg/L are toxic for some fish, so its use for treating fish disease requires careful monitoring."[111]

"Copper levels in drinking water over 1.0 mg/L result in a metallic taste and also cause blue-green staining on fixtures."[112]

"Copper levels above 1.3 mg/L can cause health related problems."[113]

"Some people who drink water containing copper in excess of the action level may, with short term exposure, experience gastrointestinal distress, and with long-term exposure may experience liver or kidney damage."[114]

"People with Wilson's Disease should consult their personal doctor if the amount of copper in their water exceeds the action level."[115]

"Inorganic chemicals include heavy metals like cadmium, copper and lead, and other compounds such as nitrate from fertilizer and asbestos."[116]

"Cyanide forms ionic complexes of varying stability with many metals. Most cyanide complexes are much less toxic than cyanide, but weak acid dissociable complexes such as those of copper and zinc are relatively unstable and will release cyanide back to the environment. Iron cyanide complexes are of particular importance due to the abundance of iron typically available in soils and the extreme stability of this complex under most environmental conditions. However, iron cyanides are subject to photochemical decomposition and will release cyanide if exposed to ultraviolet light. Metal cyanide complexes are also subject to other reactions that reduce cyanide concentrations in the environment, as described below."[117]

Iron (Fe) is cited at 0.14 mg/L in **Figure 9**.

Iron in drinking water is a very common problem.

"It can enter a water system by leaching natural deposits and from iron-bearing industrial wastes, effluents from pickling operations or acidic mine drainage."[118]

"Iron can do great economic damage when found in domestic water supplies."[119]

"Iron levels over 0.3 mg/L cause several problems."[120]

"It leaves reddish brown stains on laundry, porcelain fixtures, sinks and tubs."[121]

"It also results in a metallic taste in the water."[122]

"Higher levels of iron may also discolor the water or result in sediment."[123]

"The EPA recommended limit is 0.3 mg/L."[124]

As further cited, "Iron can be a troublesome chemical in water supplies. Making up at least 5 percent of the earth's crust, iron is one of the earth's most plentiful resources. Rainwater as it infiltrates the soil and underlying geologic formations dissolves iron, causing it to seep into aquifers that serve as sources of groundwater for wells. Although present in drinking water, iron is seldom found at concentrations greater than 10 milligrams per liter (mg/L) or 10 parts per million. However, as little as 0.3 mg/l can cause water to turn a reddish brown color."[125]

"Iron is mainly present in water in two forms: either the soluble ferrous iron or the insoluble ferric iron. Water containing ferrous iron is clear and colorless because the iron is completely dissolved. When exposed to air in the pressure tank or atmosphere, the water turns cloudy and a reddish brown substance begins to form. This sediment is the oxidized or ferric form of iron that will not dissolve in water."[126]

"Iron is not hazardous to health, but it is considered a secondary or aesthetic contaminant. Essential for good health, iron helps transport oxygen in the blood. Most tap water in the United States supplies approximately 5 percent of the dietary requirement for iron."[127]

"Dissolved ferrous iron gives water a disagreeable metallic taste. When the iron combines with tea, coffee and other beverages, it produces an inky, black appearance and a harsh, unacceptable taste. Vegetables cooked in water containing excessive iron turn dark and look unappealing."[128]

"Concentrations of iron as low as 0.3 mg/L will leave reddish brown stains on fixtures, tableware and laundry that are very hard to remove. When these deposits break loose from water piping, rusty water will flow through the faucet."[129]

"When iron exists along with certain kinds of bacteria, a smelly biofilm can form. To survive, the bacteria use the iron, leaving behind a reddish brown or yellow slime that can clog plumbing and cause an offensive odor. This slime or sludge is noticeable in the toilet tank when the lid is removed. The organisms occur naturally in shallow soils and groundwater, and they may be introduced into a well or water system when it is constructed or repaired."[130]

"Iron can combine with different naturally-occurring organic acids or tannins. Organic iron occurs when iron combines with an organic acid. Water with this type of iron is usually yellow or brown, but may be colorless. As natural organics produced by vegetation, tannins can stain water a tea color. In coffee or tea, tannins produce a brown color and react with iron to form a black residue. Organic iron and tannins are more frequently found in shallow wells, or wells under the influence of surface water."[131]

"Certain inorganic substances, such as sulfides, sulfites, thiosulfates, nitrites and ferrous iron are oxidized by dichromate, creating an inorganic COD, which is misleading when estimating the organic content of the wastewater."[132]

"ClO2 is capable of oxidizing iron and manganese, removing color, and lowering THM (Trihalomethanes) formation potential."[133]

"Cyanide forms ionic complexes of varying stability with many metals. Most cyanide complexes are much less toxic than cyanide, but weak acid dissociable complexes such as those of copper and zinc are relatively unstable and will release cyanide back to the environment. Iron cyanide complexes are of particular importance due to the abundance of iron typically available in soils and the extreme stability of this complex under most environmental conditions. However, iron cyanides are subject to photochemical decomposition and will release cyanide if exposed to ultraviolet light. Metal cyanide complexes are also subject to other reactions that reduce cyanide concentrations in the environment, as described below."[134]

"Iron cyanide complexes form insoluble precipitates with iron, copper, nickel, manganese, lead, zinc, cadmium, tin and silver. Iron cyanide forms precipitates with iron, copper, magnesium, cadmium and zinc over a pH range of 2-11."[135]

"Cyanide and cyanide-metal complexes are adsorbed on organic and inorganic constituents in soil, including oxides of aluminum, iron and manganese, certain types of clays, feldspars and organic carbon. Although the strength of cyanide retention on inorganic materials is unclear, cyanide is strongly bound to organic matter."[136]

Magnesium (Mg) is cited at 6.63 mg/L in **Figure 9**.

As reported, "Sulfates occur naturally in groundwater combined with calcium, magnesium and sodium as sulfate salts."[137]

As further cited, "Dutch drinking water contains between 1 and 5 mg of magnesium per liter."[138]

"Magnesium is present in seawater in amounts of about 1300 ppm. After sodium, it is the most commonly found cation in oceans."[139]

"A large number of minerals contains magnesium, for example dolomite (calcium magnesium carbonate; $CaMg(CO3)2$) and magnesite (magnesium carbonate; $MgCO3$)."[140]

"Magnesium is washed from rocks and subsequently ends up in water. Magnesium has many different purposes and consequently may end up in water in many different ways. Chemical industries add magnesium to plastics and other materials as a fire protection measure or as a filler. It also ends up in the environment from fertilizer application and from cattle feed. Magnesium sulphate is applied in beer breweries, and magnesium hydroxide is applied as a flocculant in wastewater treatment plants."[141]

"Magnesium is a dietary mineral for any organism but insects. It is a central atom of the chlorophyll molecule, and is therefore a requirement for plant photosynthesis."[142]

"The human body contains about 25 g of magnesium, of which 60% is present in the bones and 40% is present in muscles and other tissue. It is a dietary mineral for humans, one of the micro elements that are responsible for membrane function, nerve stimulant transmission, muscle contraction, protein construction and DNA replication. Magnesium is an ingredient of many enzymes. Magnesium and calcium often perform the same functions within the human body and are generally antagonistic."[143]

"There are no known cases of magnesium poisoning; Guidelines for magnesium content in drinking water are unlikely, because negative human and animal health effects are not expected."[144]

Manganese (Mn) is cited at 0.0330 mg/L in **Figure 9**.

As cited, "Manganese occurs naturally in many surface water and groundwater sources and in soils that may erode into these waters. However, human activities are also responsible for much of the manganese contamination in water in some areas."[145]

"A survey of snow samples near an urban expressway in Montreal, Canada (where MMT is used in petrol), was unable to establish an association between automobile emissions and manganese concentrations in the snow (Loranger et al., 1996). Loranger et al. (1994) found ambient manganese concentrations to be significantly correlated with traffic density. Areas of intermediate and high traffic densities in Montreal had ambient manganese concentrations above the natural background level of 40 ng/m3 (Loranger & Zayed, 1994; Loranger et al., 1994)."[146]

"Ambient manganese concentrations in seawater have been reported to range from 0.4 to 10 µg/l (ATSDR, 2000), with an average of about 2 µg/l (Barceloux, 1999)."[147]

"Levels in fresh water typically range from 1 to 200 µg/l (Barceloux, 1999)."

"ATSDR (2000) reported that a river water survey in the USA found dissolved manganese levels ranging from <11 to >51 µg/l."[148]

"The United States Geological Survey's National Water Quality Assessment Program has gathered limited data since 1991 on representative study basins around the USA. These data indicate a median manganese level of 16 µg/l in surface waters, with 99th-percentile concentrations of 400–800 µg/l (Leahy & Thompson, 1994; USGS, 2001)."[149]

"Higher levels in aerobic waters are usually associated with industrial pollution."[150]

"Overall, the detection frequency of manganese in groundwater in the USA is high (approximately 70% of sites) due to the ubiquity of manganese in soil and rock, but the levels detected in groundwater are generally below levels of public health concern (USEPA, 2002)."[151]

In a recent study, a "significant deficits in the intelligence quotient (IQ) of children exposed to higher concentration of manganese in drinking water" was found [see **Appendix 14**, cited below].

In addition, "at concentrations as low as 0.02 mg/l, manganese can form coatings on water pipes that may later slough off as a black precipitate (Bean, 1974). A number of countries have set standards for manganese of 0.05 mg/l, above which problems with discoloration may occur."[152]

"An epidemiological study in Japan described adverse effects in humans consuming manganese dissolved in drinking-water, probably at a concentration close to 28 mg/l (Kawamura et al., 1941)."[153]

Potassium (K) is cited at 1.48 mg/L in **Figure 9**.

As cited, "the chemistry of potassium is almost entirely that of the potassium ion, K+. The name is derived from the english word potash. The chemical symbol K comes from kalium, the Mediaeval Latin for potash, which may have derived from the arabic word qali, meaning alkali."[154]

"Most potassium occurs in the Earth's crust as minerals, such as feldspars and clays. Potassium is leached from these by weathering, which explains why there is quite a lot of this element in the sea (0.75 g/liter)."[155]

As further noted, "how old a soil is usually determines how much clay it has. The more rainfall a soil gets, the faster it breaks down into clay."[156]

"Arid regions are mostly sandy and rocky soil, unless they have areas of 'fossil' clay. River bottoms in arid regions will often have more clay because the small clay particles wash away easily from areas without vegetation cover."[157]

"Newly formed clays will often be made up of layers of silica and alumina sandwiched with potassium or iron."[158]

The Clay or Illite Group is "basically a hydrated microscopic muscovite."[159]

"The mineral illite is the only common mineral represented, however it is a significant rock forming mineral being a main component of shales and other argillaceous rocks."[160]

"The general formula is $(K, H)Al_2(Si, Al)_4O_{10}(OH)_2 - xH_2O$, where x represents the variable amount of water that this group could contain"[161] [see **Figure 14**, appearing below].

"The structure of this group is similar to the montmorillonite group with silicate layers sandwiching a gibbsite-like layer in between, in an s-g-s stacking sequence."[162]

"The variable amounts of water molecules would lie between the s-g-s sandwiches as well as the potassium ions."[163]

As further cited, "Illite is essentially a group name for non-expanding, clay-sized, dioctahedral, micaceous minerals."[164]

"It is structurally similar to muscovite in that its basic unit is a layer composed of two inward-pointing silica tetragonal sheets with a central octahedral sheet. However, illite has on average slightly more Si, Mg, Fe, and water and slightly less tetrahedral Al and interlayer K than muscovite."[165]

Figure 14: Illite Group

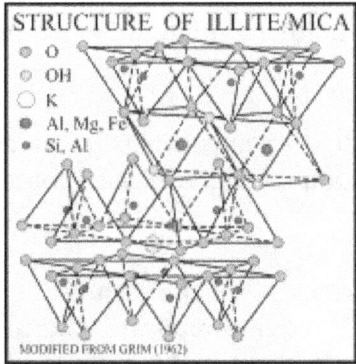

Reference to: *Illite Group*
U. S. Geological Survey Open-File Report 01-041
A Laboratory Manual for X-Ray Powder Diffraction
U.S. Department of the Interior, U.S. Geological Survey
URL: http://pubs.usgs.gov/of/2001/of01-041/htmldocs/clays/illite.htm
Maintained by Publishing Services
Last modified: 09:28:28 Thu 11 Oct 2001
http://pubs.usgs.gov/of/2001/of01-041/htmldocs/clays/illite.htm

"Together with nitrogen and phosphorous, potassium is one of the essential macrominerals for plant survival. Its presence is of great importance for soil health, plant growth and animal nutrition."[166]

"Its primary function in the plant is its role in the maintenance of osmotic pressure and cell size, thereby influencing photosynthesis and energy production as well as stomatal opening and carbon dioxide supply, plant turgor and translocation of nutrients. As such, the element is required in relatively large proportions by the growing plant."[167]

"Potassium can be found in vegetables, fruit, potatoes, meat, bread, milk and nuts. It plays an important role in the physical fluid system of humans and it assists nerve functions."[168]

"Potassium, as the ion K+, concnetrate inside cells, and 95% of the body's potassium is so located. When our kidneys are somehow malfunctioning an accumulation of potassium will consist. This can lead to disturbing heartbeats."[169]

And, "the primary inorganic nitrates which may contaminate drinking water are potassium nitrate and ammonium nitrate both of which are widely used as fertilizers."[170]

Sodium (Na) is cited at 34.3 mg/L in **Figure 9**.

As noted, "Sulfates occur naturally in groundwater combined with calcium, magnesium and sodium as sulfate salts."[171]

"Sodium is not considered a toxic element. The human body needs sodium in order to maintain blood pressure, control fluid levels and for normal nerve and muscle function."[172]

For example, "the Ontario Drinking Water Systems Regulation 170/03 under the Safe Drinking Water Act 2002 requires reporting to the local Medical Officer of Health when sodium levels in public drinking water supplies exceed 20 mg/L."[173]

"The aesthetic objective for sodium in drinking water is ≤200 mg/L. The taste of drinking water is generally considered offensive at sodium concentrations above the aesthetic objective."[174]

"Water containing more than 20 mg/L of sodium should not be used for drinking by people of severely restricted sodium diets."[175]

"Water containing more than 270 mg/L of sodium should not be used for drinking by people on moderately restricted sodium diets."[176]

Uranium (U) is cited as < 0.00050 mg/L in **Figure 9**.

As cited, "Uranium is a naturally-occurring element found at low levels in virtually all rock, soil, and water."[176]

"Significant concentrations of uranium occur in some substances such as phosphate rock deposits, and minerals such as uraninite in uranium-rich ores."[177]

"Because uranium has such a long radioactive half-life (4.47x109 years for U-238), the total amount of it on earth stays almost the same."[178]

"Uranium in soil and rocks is distributed throughout the environment by wind, rain and geologic processes. Rocks weather and break down to form soil, and soil can be washed by water and blown by wind, moving uranium into streams and lakes, and ultimately settling out and reforming as rock."[179]

"A person can be exposed to uranium by inhaling dust in air, or ingesting water and food. The general population is exposed to uranium primarily through food and water. The average daily intake of uranium from food ranges from 0.07 to 1.1 micrograms per day."[180]

"About 99 percent of the uranium ingested in food or water will leave a person's body in the feces, and the remainder will enter the blood. Most of this absorbed uranium will be removed by the kidneys and excreted in the urine within a few days. A small amount of the uranium in the bloodstream will deposit in a person's bones, where it will remain for years."[181]

"EPA standards under the Clean Air Act limit uranium in the air. The maximum dose to an individual from uranium in the air is 10 millirem."[182]

Zinc (Zn) is cited as < 0.020 mg/L in **Figure 9**.

As noted, "in natural surface waters, the concentration of zinc is usually below 10 μg/litre, and in ground waters, 10–40 μg/litre."[183]

"In tap water, the zinc concentration can be much higher as a result of the leaching of zinc from piping and fittings."[184]

As further cited, "Zinc is commonly found in many natural waters."[185]

"The deterioration of galvanized iron and leaching of brass can add substantial amounts of zinc to water."[186]

"Industrial effluents may also contribute large amounts of zinc to drinking water."[187]

"Zinc is essential to human metabolism and has been found to be necessary for proper body growth."[188]

"Although essential in our diet, high zinc concentrations in water can irritate the human digestive system."[189]

"Levels above 5 mg/L cause a bitter metallic taste and opalescence in alkaline drinking water."[190]

"High concentrations of zinc suggest the presence of lead and cadmium, common impurities from the galvanizing process."[191]

"The EPA recommended limit is 5 mg/L."[192]

Ammonia as N is cited as < 0.050 mg/L in **Figure 9**.

As noted, "several water properties and constituents affect the efficiency of chlorination, and therefore affect the amount of chlorine that must be added to achieve levels of chlorine compounds that have disinfecting capability (hypochlorous acid, hypochlorite and inorganic chloramine)."[193]

"Materials that react with chlorine and reduce or eliminate the disinfecting ability include ammonia, ammonium, organics and reducing agents."[194]

"The 'interfering' compounds reduce the disinfecting ability of added chlorine, and as their level increases, the amount of chlorine that must be added increases."[195]

"At high levels of interfering compounds, however, it can be difficult for water treatment facilities to add enough chlorine to provide satisfactory levels of disinfecting compounds in the water and have reactions proceed at a rapid enough pace. It also adds costs for treatment."[196]

As further cited, "Ammonia is very soluble in water; approximately 90 g dissolve in 100 mL of distilled water at 0°C."[197]

"In solution, some of the ammonia reacts with the water, resulting in the following equilibrium:

$$NH_3 + H_2O \rightarrow NH_4^+ + OH^-, pK_b = 4.74$$"[198]

"The concentration of ammonia in Canadian surface waters in a 1980-1981 survey ranged from <0.001 to 0.587 mg/L."[199]

"In a survey of 19 public water supplies conducted in Ontario in 1985 and 1986, it was found that the mean level of ammonia (as nitrogen) in untreated water was 0.20 mg/L, with a range of <0.02-0.65 mg/L."[200]

"After treatment, the mean level was 0.17 mg/L (range <0.02-0.40 mg/L), whereas the mean level in the distribution system was 0.15 mg/L (range <0.02-0.44 mg/L)."[201]

We have cited above, the results of the water test before the raw sewage spill [see **Figure 9**, cited below].

We will use these results as a baseline to compare to the results of the water test after the raw sewage spill, as cited in the next chapter.

Figure 9: Water quality before the raw sewage spill

Sample Details/Parameters	Result	Qualifier*	D.L.	Units	Extracted
Total Coliform, EColi Mcoli Blue & HPC					
Escherichia Coli mcoli blue MF					
E. Coli	<1		1	CFU/100mL	10-AUG-11
Heterotrophic Plate Count					
Heterotrophic Plate Count	<10		10	CFU/mL	10-AUG-11
Total Coliform mcoli blue MF					
Total Coliforms	<1		1	CFU/100mL	10-AUG-11
WP2 Drinking Water					
Chloride					
Chloride	23.4		0.50	mg/L	
Conductivity					
Conductivity	293		0.40	umhos/cm	
Fluoride					
Fluoride	0.70		0.10	mg/L	
Hardness - grains/Imperial gallon					
Hardness-grains IMPgal	6.26		0.010	grn/IMPgal	
Hardness - grains/US gallon					
Hardness-grains/USgal	5.22		0.010	grn/USgal	
Hardness Calculated					
Hardness (as CaCO3)	89.2		0.30	mg/L	
Nitrate as N					
Nitrate-N	<0.050		0.050	mg/L	
Nitrate+Nitrite					
Nitrate and Nitrite as N	<0.071		0.071	mg/L	
Nitrite as N					
Nitrite-N	<0.050		0.050	mg/L	
Sulfate					
Sulfate	55.1		0.50	mg/L	
TDS (Calculated from EC)					
TDS (Calculated from EC)	190		20	mg/L	
Total Metals by ICP-MS					
Arsenic (As)-Total	<0.0010		0.0010	mg/L	11-AUG-11
Barium (Ba)-Total	0.0151		0.00050	mg/L	11-AUG-11
Boron (B)-Total	<0.030		0.030	mg/L	11-AUG-11
Calcium (Ca)-Total	24.8		0.20	mg/L	11-AUG-11
Copper (Cu)-Total	0.0341		0.0020	mg/L	11-AUG-11
Iron (Fe)-Total	0.14		0.10	mg/L	11-AUG-11
Magnesium (Mg)-Total	6.63		0.050	mg/L	11-AUG-11
Manganese (Mn)-Total	0.0330		0.0010	mg/L	11-AUG-11
Potassium (K)-Total	1.48		0.10	mg/L	11-AUG-11
Sodium (Na)-Total	34.3		0.050	mg/L	11-AUG-11
Uranium (U)-Total	<0.00050		0.00050	mg/L	11-AUG-11
Zinc (Zn)-Total	<0.020		0.020	mg/L	11-AUG-11
pH					
pH	7.98		0.10	pH units	
Miscellaneous Parameters					
Ammonia as N	<0.050		0.050	mg/L	
Phosphorus (P)-Total	0.612		0.010	mg/L	

Footnotes

1 - 2. *Partially treated sewage flows from Winnipeg plant*
CBC News
Posted: Nov 2, 2011 4:36 PM CT
http://www.cbc.ca/news/canada/manitoba/story/2011/11/02/winnipeg-wastewater-plant-problem.html

3 - 4. *Sewage treatment plant spewing into Red*
By Paul Turenne, Winnipeg Sun
First posted: Wednesday, November 02, 2011
http://www.winnipegsun.com/2011/11/02/sewage-treatment-plant-spewing-into-red

5 - 8. Make sewers election issue
By Tom Brodbeck,Winnipeg Sun
 First posted: Monday, June 06, 2011 10:46 PM CDT
http://www.winnipegsun.com/2011/06/06/make-sewers-election-issue

9. *Escherichia coli: Bacteriological guidelines*
http://www.hc-sc.gc.ca/ewh-semt/pubs/water-eau/2010-sum_guide-res_recom/index-eng.php

Also see: *No observable adverse effect level*
http://en.wikipedia.org/wiki/No_observable_adverse_effect_level

"The no observable adverse effect level (NOAEL) denotes the level of exposure of an organism, found by experiment or observation, at which there is no biologically or statistically significant (e.g. alteration of morphology, functional capacity, growth, development or life span) increase in the frequency or severity of any adverse effects in the exposed population when compared to its appropriate control.

In toxicology it is specifically the highest tested dose or concentration of a substance (i.e. a drug or chemical) or agent (e.g. radiation), at which no such adverse effect is found in exposed test organisms where higher doses or concentrations resulted in an adverse effect.

This level may be used in the process of establishing a dose-response relationship, a fundamental step in most risk assessment methodologies.

In drug development, NOAEL of a new drug is assessed in laboratory animals drugs prior to initiation of clinical trials to establish a safe clinical starting dose in human trials.

The United States Environmental Protection Agency (US EPA) defines NOAEL as 'an exposure level at which there are no statistically or biologically significant increases in the frequency or severity of adverse effects between the exposed population and its appropriate control; some effects may be produced at this level, but they are not considered as adverse, or as precursors to adverse effects. In an experiment with several NOAELs, the regulatory focus is primarily on the highest one, leading to the common usage of the term NOAEL as the highest exposure without adverse effects'."

Also see: *Lowest-observed-adverse-effect-level*

"The lowest-observed-adverse-effect level is the lowest concentration or amount of a substance found by experiment or observation that causes an adverse alteration of morphology, function, capacity, growth, development, or lifespan of a target organism distinguished from normal organisms of the same species under defined conditions of exposure. Federal agencies use set approval standards below this level."

10 - 13. *Guidelines for Canadian Drinking Water Quality: 3.2 Coliform, Coliform Background and Heterotrophic Plate Counts*
Sixth Edition
Published by authority of the Minister of Health
1996
https://docs.google.com/viewer?a=v&q=cache:OwJdLz9-vywJ:xnet.rrc.mb.ca/rcharney/water%2520quality.pdf+NOAEL+heterotrophic+plate+count+Escherichia+coli&hl=en&gl=ca&pid=bl&srcid=ADGEEShmUBJ6BCXV_LDJUfOBPwlVhfkwPrR5rK72LhfJMopt4-PfTs0h4j_mJNP41zRI6eTMdjxVL4-VSHANgJGbLFup0z_JYa7h6rUBfE556UdOlrxBk6g6hDaHTywCeiny3wyPqrF8&sig=AHIEtbTAgn52pi9Xsjm0giDBSuxmALgsLA

14 - 17. *EPA Approved Method for E. Coli Determination*
http://www.lamotte.com/component/option,com_pages/lang,en/mid,/page,244

"USEPA Approved for the determination of E. coli and total coliforms for use in National Primary Drinking Water regulations (NPDWR) monitoring.

1.1 The Coliscan® Membrane Filter method consists of a medium which detects the presence or absence of E. coli and total coliforms simultaneously and also allows the enumeration of each. It is approved for use in analyzing potable water by certified drinking water laboratories. It is also useful in the analysis of other waters, clinical applications, veterinary and agricultural applications, pharmaceutical applications and in the area of food and beverages."

18 - 20. *SCIENTIFIC COMMITTEE ON TOXICITY, ECOTOXICITY AND THE ENVIRONMENT (CSTEE), OPINION OF THE CSTEE ON "STUDIES CONCERNING THE QUALITY OF DRINKING WATER IN SELECTED EUROPEAN CITIES."*
EUROPEAN COMMISSION: HEALTH & CONSUMER PROTECTION DIRECTORATE-GENERAL
Directorate C - Scientific Opinions
C2 - Management of scientific committees; scientific co-operation and networks
Brussels, C2/VR/csteeop/ 12062003/D(03)
Opinion expressed at the 38th CSTEE plenary meeting
Brussels, 12 June 2003
https://docs.google.com/viewer?a=v&q=cache:x8XHQek61ZQJ:ec.eu ropa.eu/food/fs/sc/sct/out187_en.pdf+NOAEL+heterotrophic+plate+c ount+Escherichia+coli&hl=en&gl=ca&pid=bl&srcid=ADGEEShrng NYB_pHPI8A_SZ4044FWgX2d3BcD3J1jWCQWo- eqx0OO6oF7LCwvHK2DJZvYt4NhdMOEG_HUdNRqaDI48I2WgtWw UhkR7abaWVVy7yhnTAubJ6cNZIxxV5rrz9tI5qaAITb&sig=AHIEtbT F1q6DD91NRLG9SSX4U30l9dyeUw

21. Chris Mechenich & Elaine Andrews,
Interpreting Drinking Water Test Results
G3558 - 4
http://www4.uwsp.edu/cnr/gndwater/privatewells/Interpreting%20Dri nking%20Water%20Test%20Results.pdf

22. *Case study – County of Oxford, Ontario*
http://www.ec.gc.ca/eau-water/default.asp?lang=En&n=F33CB10C-1

23 - 32. *Water: Monitoring & Assessment: 5.9 Conductivity*
http://water.epa.gov/type/rsl/monitoring/vms59.cfm

33. *Case study – County of Oxford, Ontario*
http://www.ec.gc.ca/eau-water/default.asp?lang=En&n=F33CB10C-1

34. Chris Mechenich & Elaine Andrews,
Interpreting Drinking Water Test Results
G3558 - 4
http://www4.uwsp.edu/cnr/gndwater/privatewells/Interpreting%20Dri nking%20Water%20Test%20Results.pdf

35. *Is Your Drinking Water Safe?*
A Guide to Water Testing
Brett Hopkins
Silver Lake Research - Watersafe
March 2002

http://www.elements.nb.ca/theme/water/silverlake/testing.htm

Also see: Drinking water quality and health
by P. Kendall1(3/2010)
http://www.ext.colostate.edu/pubs/foodnut/09307.html

"Hard vs. Soft Water - The hardness of water relates to the amount of calcium, magnesium and sometimes iron in the water. The more minerals present, the harder the water. Soft water may contain sodium and other minerals or chemicals; however, it contains very little calcium, magnesium or iron. Many people prefer soft water because it makes soap lather better, gets clothes cleaner and leaves less of a ring around the tub. Some municipalities and individuals remove calcium and magnesium, both essential nutrients, and add sodium in an ion-exchange process to soften their water. The harder the water, the more sodium that must be added in exchange for calcium and magnesium ions to soften the water. This process has drawbacks from a nutritional standpoint.

First, soft water is more likely to dissolve certain metals from pipes than hard water. These metals include cadmium and lead, which are potentially toxic. Second, soft water may be a significant source of sodium for those who need to restrict their sodium intake for health reasons. Approximately 75 milligrams of sodium is added to each quart of water per 10 g.p.g. (grains per gallon) hardness. Finally, there is epidemiological evidence to suggest a lower incidence of heart disease in communities with hard water. The Environmental Protection Agency (EPA) doesn't set a mandatory upper limit for sodium in water, but suggests an upper limit of 20 milligrams per liter (quart) to protect individuals on sodium-restricted diets."

36 - 42. *Hardness in Drinking-water: Background document for development of WHO Guidelines for Drinking-water Quality* WHO/HSE/WSH/10.01/10/Rev/1
https://docs.google.com/viewer?a=v&q=cache:hTFpGVwNp3MJ:www.who.int/water_sanitation_health/dwq/chemicals/hardness.pdf+hardness+drinking+water+quality&hl=en&gl=ca&pid=bl&srcid=ADGEESgK2SoZ7ZbmsLjceSOzU8XspVCcCXuz8685JrLToqK1zv93bn_za3asnzmOrHYAcq4Vb8HPZDLbdT1aCacQzn4XTimfLJZuPLmeJ4o2VcDuWuOHXXKi8q_-YO_ujoHtmQFRurzq&sig=AHIEtbRxVn-B0-JTHSpM5ve_pCcpLabLkQ

Also see: Altman D et al. (2002) *Do women with pre-eclampsia, and their babies, benefit from magnesium sulphate? The Magpie Trial: a randomised placebo-controlled trial.* Lancet, 359(9321):1877–1890.

Also see: Cotruvo JA et al. (2010) *Desalination technology: health and environmental impacts.* Boca Raton, FL, CRC Press.

Also see: Langan SM (2009) *Flares in childhood eczema.* (http://www.skintherapyletter.com/2009/14.8/2.html).

Also see: Leurs LJ et al. (2010) *Relationship between tap water hardness, magnesium, and calcium concentration and mortality due to ischemic heart disease or stroke in the Netherlands.* Environmental Health Perspectives, 118(3):414–420.

Also see: McGowan W (2000) *Water processing: residential, commercial, light-industrial,* 3rd ed. Lisle, IL, Water Quality Association.

Also see: McNally NJ et al. (1998) *Atopic eczema and domestic water hardness.* Lancet, 352(9127):527–531.

Also see: Neri LC, Johansen HL (1978) *Water hardness and cardiovascular mortality.* Annals of the New York, Academy of Sciences, 304:203–221.

Also see: Neri LC et al. (1985) *Magnesium and certain other elements and cardiovascular disease.* Science of the Total Environment, 42:49–75.

Also see: Ong CN, Grandjean AC, Heaney RP (2009) *The mineral composition of water and its contribution to calcium and magnesium intake. In: Calcium and magnesium in drinking-water: public health significance.* Geneva, World Health Organization, pp. 36–58 (http://whqlibdoc.who.int/publications/2009/9789241563550_eng.pdf).

Also see: Thomas KS, Sach TH (2000) *A multicentre randomized controlled trial of ion-exchange water softeners for the treatment of eczema in children: protocol for the Softened Water Eczema Trial (SWET) (ISRCTN: 71423189).* British Journal of Dermatology, 159(3): 561–566.

43. *Hard water in Canada*
http://en.wikipedia.org/wiki/Hard_water#Hard_water_in_Canada

44. *Biodegradation*
Environmental and Health Effects of Cyanide
http://www.cyanidecode.org/cyanide_environmental.php

45. *Nitrates & Nitrites: Is Your Drinking Water Safe?*
A Guide to Water Testing
Brett Hopkins
Silver Lake Research - Watersafe

March 2002
http://www.elements.nb.ca/theme/water/silverlake/testing.htm

46 - 51. *Water Test, Inc. tests for the following contaminants*:
http://www.watertestinc.com/contaminants.html

52. Chris Mechenich & Elaine Andrews,
Interpreting Drinking Water Test Results
G3558 - 4
*http://www4.uwsp.edu/cnr/gndwater/privatewells/Interpreting%20Dri
nking%20Water%20Test%20Results.pdf*

53 - 56. *Drinking water quality and health*
by P. Kendall1(3/2010)
http://www.ext.colostate.edu/pubs/foodnut/09307.html

57. *CHLORINE DIOXIDE*
http://home.windstream.net/mikeric/Odor/clo2.htm

58. *Water Quality Terms*
http://www.water-research.net/glossary.htm

59. *Sulfate in Drinking Water*
http://water.epa.gov/drink/contaminants/unregulated/sulfate.cfm

60. *Water Quality Terms*
http://www.water-research.net/glossary.htm

61. *Sulfate in Drinking Water*
http://water.epa.gov/drink/contaminants/unregulated/sulfate.cfm

62 - 64. *Arsenic in Drinking Water*
http://water.epa.gov/lawsregs/rulesregs/sdwa/arsenic/index.cfm

Also see: *Arsenic Rule*
http://water.epa.gov/lawsregs/rulesregs/sdwa/arsenic/regulations.cfm

65 – 66. *Arsenic in Drinking Water*
Natural Resources Defense Council
http://www.nrdc.org/water/drinking/qarsenic.asp

67 - 72. *Arsenic*
http://www.hc-sc.gc.ca/hl-vs/iyh-vsv/environ/arsenic-eng.php

Also see: *Fairbanks, Alaska: Case Study in Arsenic Treatment Technologies*
EPA 816-F-03-012 May 2003

http://water.epa.gov/drink/info/arsenic/upload/2006_11_28_arsenic_casestudies_cas
estudy_fairbanks.pdf

Also see: *Technologies and Costs for Removal of Arsenic from Drinking Water*
EPA 815-R-00-028 December 2000
http://water.epa.gov/drink/info/arsenic/upload/2005_11_10_arsenic_treatments_and
_costs.pdf

73 - 79. *Drinking Water Contaminants- Barium*
http://www.freedrinkingwater.com/water-contamination/barium-removal-water.htm

Also see: *Basic Information about Barium in Drinking Water*
http://water.epa.gov/drink/contaminants/basicinformation/basicinformation_barium.
cfm

80 – 94. *Boron in Drinking-water: Background document for development of*
WHO Guidelines for Drinking-water Quality
WHO/SDE/WSH/03.04/54
https://docs.google.com/viewer?a=v&q=cache:_roudP7RyEgJ:www.who.int/water_
sanitation_health/dwq/boron.pdf+boron+limits+in+drinking+water&hl=en&gl=ca&
pid=bl&srcid=ADGEESgEia7Wch-
39T_1bzrLQdeWS0Wle_UZh186Y4mqTXXfTy0bmzcYY1kBWk2pk3NRrxP7qsh
-4wEhL5lPYX-NR-
g4ofhmICOmSW63ge2j9V9kRhjVO8P1cUP53abi28nnV2BCSRsQ&sig=AHIEtbR
0agJaUeU5uVo_Ltw8ucA5GIx6ZA

Also see: Budavari S et al., eds. (1989) *The Merck index*, 11th ed. Rahway, NJ,
Merck and Co., Inc.

Also see: Cotton PA, Wilkinson L (1988) *Advanced inorganic chemistry*, 5th ed.
New York, NY, John Wiley & Sons, pp. 162-165.

Also see: Rai D et al. (1986) "Chemical attenuation rates, coefficients, and
constants in leachate migration". *Vol. 1: A critical review*. Report to Electric Power
Research Institute, Palo Alto, CA, by Battelle Pacific Northwest Laboratories,
Richland, WA (Research Project 2198-1).

Also see: Waggott A (1969) *An investigation of the potential problem of increasing*
boron concentrations in rivers and water courses. Water research, 3:749-765.

Also see: Keren R, Mezuman U (1981) *Boron adsorption by clay minerals using a*
phenomenological equation. Clays and clay minerals, 29:198-204.

Also see: Keren R, Gast RG, Bar-Yosef B (1981) *pH-dependent boron adsorption*
by Na-montmorillonite. Soil Science Society of America journal, 45:45-48.
(Sprague, 1972). Mance et al. (1988)

Also see: Linder RE, Strader LF, Rehnberg GL (1990) *Effect of acute exposure to boric acid on the male reproductive system of the rat*. Journal of toxicology and environmental health, 31:133 - 146.

Also see: Weir RJ, Fisher RS (1972) *Toxicologic studies on borax and boric acid*. Toxicology and applied pharmacology, 23:351-364.

Also see: Price CJ et al. (1996a) *Developmental toxicity NOAEL and postnatal recovery in rats fed boric acid during gestation*. Fundamental and applied toxicology, 32:179-193.

Also see: Price CJ et al. (1996b) *The developmental toxicity of boric acid in rabbits*. Fundamental and applied toxicology, 34:176-187.

Also see: Heindel JJ et al. (1992) *Developmental toxicity of boric acid in mice and rats*. Fundamental and applied toxicology, 18:266-277.

95. *Ambient Water Quality Guidelines for Boron: Overview Report: Table 1: Recommended Guidelines for Boron*
Prepared pursuant to Section 2(e) of the Environment Management Act, 1981
Original Signed by Margaret Eckenfelder, Assistant Deputy Minister
Water, Land and air Protection, July 23, 2003
http://www.env.gov.bc.ca/wat/wq/BCguidelines/boron/boron.html

97. *Drinking water quality and health*
by P. Kendall1(3/2010)
http://www.ext.colostate.edu/pubs/foodnut/09307.html

98. *Is Your Drinking Water Safe?*
A Guide to Water Testing
Brett Hopkins
Silver Lake Research - Watersafe
March 2002
http://www.elements.nb.ca/theme/water/silverlake/testing.htm

99. *Calcium In Drinking Water*
http://www.our-drinking-water.com/calcium-in-drinking-water.html

100 - 101. Simon Morr, Esteban Cuartas, Basil Alwattar, and Joseph M. Lane
How Much Calcium Is in Your Drinking Water? A Survey of Calcium Concentrations in Bottled and Tap Water and Their Significance for Medical Treatment and Drug Administration
HSS J. 2006 September; 2(2): 130–135
http://www.ncbi.nlm.nih.gov/pmc/articles/PMC2488164/

102. *Drinking water quality and health*
by P. Kendall1(3/2010)
http://www.ext.colostate.edu/pubs/foodnut/09307.html

103 - 106. *Basic Information about Copper in Drinking Water*
http://water.epa.gov/drink/contaminants/basicinformation/copper.cfm

107 – 113. *Water Test, Inc. tests for the following contaminants*:
http://www.watertestinc.com/contaminants.html

114 - 115. *Basic Information about Copper in Drinking Water*
http://water.epa.gov/drink/contaminants/basicinformation/copper.cfm

116. *Toxic Water Pollutants*
By John Marton, Jun 8, 2010
http://www.livestrong.com/article/142654-toxic-water-pollutants/

117. Reference to: *Environmental and Health Effects of Cyanide*
http://www.cyanidecode.org/cyanide_environmental.php

118 – 124. *Water Test, Inc. tests for the following contaminants*:
http://www.watertestinc.com/contaminants.html

125 - 131. *Iron In Drinking Water*
http://www.idph.state.il.us/envhealth/factsheets/ironFS.htm

132 – 133. *CHLORINE DIOXIDE*
http://home.windstream.net/mikeric/Odor/clo2.htm

134 – 136. Reference to: *Environmental and Health Effects of Cyanide*
http://www.cyanidecode.org/cyanide_environmental.php

137. *Drinking water quality and health*
by P. Kendall1(3/2010)
http://www.ext.colostate.edu/pubs/foodnut/09307.html

138 - 144. *Magnesium (Mg) and water: reaction mechanisms, environmental impact and health effects*
http://www.lenntech.com/periodic/water/magnesium/magnesium-and-water.htm

145 - 153. *Manganese in Drinking-water: Background document for development of WHO Guidelines for Drinking-water Quality*
WHO/SDE/WSH/03.04/104/Rev/1
https://docs.google.com/viewer?a=v&q=cache:ITUAq5eXxPYJ:www.who.int/water_sanitation_health/dwq/chemicals/manganese.pdf+NOAEL+manganese+drinking+water&hl=en&gl=ca&pid=bl&srcid=ADGEESjSQcOImhIhyJ2U7jJ91dk2VyxiuUjI

Z1v3FAvKOdMGzXWjJBl9gnayGcfQv_QP0WPc8b4Gp0K6UReWbZQpfu8yzIhP
adQeeZ1f2R_UvVKJ-srJo7cZ-
2pu7nsgfgmxDkBVqlbB&sig=AHIEtbQh7_lXRscjr9zMeRzIjA-JsE2ZSg

Also see: *ATSDR (2000) Toxicological profile for manganese.* Atlanta, GA, United States Department of Health and Human Services, Public Health Service, Agency for Toxic Substances and Disease Registry.

Also see: Barceloux DG (1999) *Manganese.* Clinical Toxicology, 37:293–307.

Also see: Bean EL (1974) *Potable water quality goals.* Journal of the American Water Works Association, 66:221.

Also see: Kawamura CL et al. (1941) *Intoxication by manganese in well water.* Kitasato Archives of Experimental Medicine, 18:145–169.

Also see: Leahy PP, Thompson TH (1994) *The National Water-Quality Assessment Program.* Washington, DC, United States Geological Survey, 4 pp. Available at http://water.usgs.gov/nawqa/NAWQA.OFR94-70.html (Open-File Report 94-70).

Also see: Loranger S, Zayed J (1994) *Manganese and lead concentrations in ambient air and emission rates from unleaded and leaded gasoline between 1981 and 1992 in Canada: A comparative study.* Atmospheric Environment, 28:1645–1651.

Also see: Loranger S, Zayed J (1995) *Environmental and occupational exposure to manganese: A multimedia assessment.* International Archives of Occupational and Environmental Health, 67:101–110.

Also see: Loranger S, Zayed J, Forget E (1994) *Manganese contamination in Montreal in relation with traffic density.* Water, Air and Soil Pollution, 74:385–396.

Also see: Loranger S et al. (1996) *Manganese and other trace elements in urban snow near an expressway.* Environmental Pollution, 92:203–211.

Also see: USGS (2001) *US Geological Survey National Water Quality Assessment Data Warehouse.* Available at http://infotrek.er.usgs.gov/pls/nawqa/nawqa.home.

Also see: USEPA (1984) *Health assessment document for manganese.* Cincinnati, OH, United States Environmental Protection Agency, Environmental Criteria and Assessment Office (EPA-600/8-83-013F).

Also see: USEPA (1990) *Comments on the use of methylcyclopentadienyl manganese tricarbonyl in unleaded gasoline.* Research Triangle Park, NC, United States Environmental Protection Agency, Office of Research and Development.

Also see: USEPA (1994) *Drinking water criteria document for manganese.* Washington, DC, United States Environmental Protection Agency, Office of Water (September 1993; updated March 1994).

Also see: USEPA (1996) *Exposure factors handbook. Vol. 1. General factors.* Washington, DC, United States Environmental Protection Agency (EPA/600/8-89/043).

Also see: USEPA (1997) *Manganese.* Washington, DC, United States Environmental Protection Agency, Integrated Risk Information System (IRIS). Available at http://www.epa.gov/iris/subst/0373.htm.

Also see: USEPA (2002) *Health effects support document for manganese.* Washington, DC, United States Environmental Protection Agency, Office of Water.

154 - 155. *Potassium - K*
http://www.lenntech.com/periodic/elements/k.htm

156 – 158. *Cation Exchange Capacity in Soils, Simplified*
http://www.soilminerals.com/Cation_Exchange_Simplified.htm

159 – 163. *The Clay Minerals*
http://www.galleries.com/Clays_Group

164 - 165. *Illite Group*
U. S. Geological Survey Open-File Report 01-041
A Laboratory Manual for X-Ray Powder Diffraction
U.S. Department of the Interior, U.S. Geological Survey
URL: http://pubs.usgs.gov/of/2001/of01-041/htmldocs/clays/illite.htm
Maintained by Publishing Services
Last modified: 09:28:28 Thu 11 Oct 2001
http://pubs.usgs.gov/of/2001/of01-041/htmldocs/clays/illite.htm

166 – 169. *Potassium - K*
http://www.lenntech.com/periodic/elements/k.htm

170. *Drinking Water Contaminants - Nitrates / Nitrites*
http://www.freedrinkingwater.com/water-contamination/nitrates-nitrites-contaminants-removal-water.htm

171. *Drinking water quality and health*
by P. Kendall1(3/2010)
http://www.ext.colostate.edu/pubs/foodnut/09307.html

172 - 174. Sodium in drinking water: Fact Sheet

https://docs.google.com/viewer?a=v&q=cache:NDezDYnRO0IJ:www.townofstmar
ys.com/uploadedFiles/Town_Services/Water_Department/Sodium%2520Flyer.pdf+
Sodium+maximum+level+drinking+water&hl=en&gl=ca&pid=bl&srcid=ADGEES
j35htH_Vr_tptqE-BthempEb3IOnas7sv98HRKAh1YQHKMPV30LDrunRvS-
qXKsRWiksRkbfv5o8QnPFRzUSbBS_IxVt86CbXdiPLFbPJMx4vNwceknD0ShK
LwKMKjgeCd3-Xi&sig=AHIEtbQAJm6bVlFt-nBjC2JF5z6FEk7pvg

175 – 176. *Table of Drinking Water Contaminant Levels - New York State
Maximum Contamination Allowed in Water*
http://www.inspectapedia.com/water/levels.htm

177 - 182. *Uranium*
http://www.epa.gov/radiation/radionuclides/uranium.html

183 – 184. *Zinc in Drinking-water*
Background document for development of WHO Guidelines for Drinking-water
Quality
WHO/SDE/WSH/03.04/17
https://docs.google.com/viewer?a=v&q=cache:B6T3anolMacJ:www.who.int/water_
sanitation_health/dwq/chemicals/zinc.pdf+Zinc+maximum+level+drinking+water&
hl=en&gl=ca&pid=bl&srcid=ADGEESio9HNjGtp5JrXJEtvQFmqDK11igSly6_stL
Y3AOmBrtZbaxVsTtLMYACaXmeo7Rj57bExZLcCTMEK1CZvEk7XegZbb8Pu7
PQ2Z4yR8CjytpSq9Xd3E8yJkRzqB3nDnZC2cygjj&sig=AHIEtbSTUjS23V84-
WuNIn0FZ0AgFL8gKw

Also see: Elinder CG. *Zinc*. In: Friberg L, Nordberg GF, Vouk VB, eds. Handbook
on the toxicology of metals, 2nd ed. Amsterdam, Elsevier Science Publishers,
1986:664-679.

Also see: Nriagu JO, ed. *Zinc in the environment. Part I, Ecological cycling.* New
York, NY, John Wiley, 1980.

185 – 192. *Water Test, Inc. tests for the following contaminants*:
http://www.watertestinc.com/contaminants.html

193 - 196. *Surface Waters: Ammonium is Not Ammonia – Part Two*
By John Sawyer, Department of Agronomy
http://www.extension.iastate.edu/CropNews/2008/0502JohnSawyer.htm

197 - 201. *Ammonia*
Environmental and Workplace Health
http://www.hc-sc.gc.ca/ewh-semt/pubs/water-eau/ammonia-ammoni/index-eng.php

Appendix 10

EPA Approved Method for E. Coli Determination
http://www.lamotte.com/component/option,com_pages/lang,en/mid,/pa
ge,244

The COLISCAN® MEMBRANE FILTER METHOD Developed for
ESCHERICHIA COLI AND TOTAL COLIFORMS

USEPA Approved for the determination of E. coli and total coliforms
for use in National Primary Drinking Water regulations (NPDWR)
monitoring.
1.0 Scope and Application

1.1 The Coliscan® Membrane Filter method consists of a medium
which detects the presence or absence of E. coli and total coliforms
simultaneously and also allows the enumeration of each. It is approved
for use in analyzing potable water by certified drinking water
laboratories. It is also useful in the analysis of other waters, clinical
applications, veterinary and agricultural applications, pharmaceutical
applications and in the area of food and beverages.

1.2 The method allows the detection and enumeration of E. coli and
other coliforms in 24 hours or less and does not require a confirmation
step.

1.3 The detection limit is one target CFU/sample.

2.0 Summary of the Method

2.1 The Coliscan® MF medium determines the presence or absence
(and enumeration) of E. coli and other coliforms in any size water
sample (100 mL is required for drinking water). The sample (diluted or
not) is passed through a 0.45um pore size, 47 mm diameter membrane
filter using standard equipment and methodology. The filter is then
placed into a 50 mm plate containing a pad saturated with the medium
(if in broth form) or a layer of the medium which has been solidified
with an added agar gelling agent. Incubate for 24 hr at 35°C±0.5°C. E.
coli CFUs will appear as blue/purple and other coliform CFUs will
appear pink/magenta.

2.2 Coliscan® MF medium (broth or agar) contains nutrients to assure
the growth of the target organisms, buffers to maintain appropriate pH,
and inhibitors to reduce the growth of nontarget organisms. It is similar
to the modification of m-TEC described by Duncanson and Cabelli
(1986 paper presented at the National Meeting of AWWA). E. coli

colonies growing on the medium appear blue to purple due to the combination of the enzymes glucuronidase and galactosidase affecting their respective substrates, 5-Bromo-4-Chloro-3-Indolyl-B-D-glucuronide (X-gluc) and 6-Chloro-3-Indolyl-B-D-galactoside (Red-Gal®). The teal green product of X-gluc hydrolysis combines with the pink/magenta product of the Red-Gal® hydrolysis to produce the blue to purple appearance of the E. coli colonies. Other coliform colonies (than E. coli) are colored pink/magenta as a result of producing only the galactosidase which acts on the Red-Gal® only.

3.0 Definitions

3.1 Escherichia coli - Those bacteria which grow as blue/purple colonies on the Coliscan® MF medium as a result of the production of both glucuronidase and galactosidase enzymes. These bacteria are of fecal origin.

3.2 Total Coliforms - Those bacteria which make up the sum of the E. coli (blue/purple colonies) and other coliforms. The other coliforms will appear as pink/magenta colonies on the Coliscan® MF medium because they produce galactosidase, but not glucuronidase and so cleave only the Red-Gal® substrate. Species of the genera Citrobacter, Enterobacter, Escherichia, and Klebsiella are the main groups (other than Escherichia) of coliform bacteria.

3.3 Non Coliforms - Bacteria that form colonies which are not blue/purple or pink/magenta on Coliscan® MF medium.

4.0 Interferences

4.1 No known chemical substances normally encountered in drinking water or source waters have been observed to affect the color of E. coli or other coliform colonies on the Coliscan® MF medium. If particulate or colloidal materials are suspended in water samples, they may interfere with filtering efficiency by clogging filter pores and they may cause some spreading of bacterial colonies as they grow on the filter surface during incubation. However these materials would very rarely prevent the accurate determination of the bacterial population.

4.2 Colonies exhibiting the color(s) of the target organisms should not be included in the E. coli or other coliform counts if they are less than 0.5 mm diameter (except when the entire colony population is small due to excessive crowding on the plate. In such a case the sample should be rerun at a higher dilution.). Small colored colonies of this nature should not be counted unless they are isolated into pure culture and then verified by approved procedures.

4.3 Colonies exhibiting a teal green color which is indicative of the production of glucuronidase without the production of galactosidase should not be counted as E. coli without isolation into pure culture and verification by approved procedures.

4.4 It can not be safely assumed that colonies can be picked directly from the surface of the filter and used to inoculate confirmatory media directly, particularly if the colonies are blue/purple or teal green, as they may be contaminated by cells from adjoining colonies that have traveled on the filter surface. Therefore, questionable colonies should be picked and streaked onto the surface of a differential medium such as Coliscan® Easygel® to ensure their purity before further testing.

4.5 Unlike media which utilize a fluorogen (such as MUG or MUGal) and a chromogen, where the fluorogen tends to diffuse rapidly into the surrounding medium, thus making the timing of reading the test results critical (before excessive diffusion occurs which may make neighboring colonies appear as false positives), the chromophores of the chromogens used in Coliscan® MF are quite insoluble and little or no diffusion away from the target colonies occurs.

5.0 Safety

5.1 Standard safety practices should be observed by persons using these materials in the laboratory.

5.2 Any materials containing living or viable microbes should be disinfected or sterilized by accepted standard methods before being discarded.

5.3. Refer to the MSDS for specific product information.

6.0 Equipment and supplies

6.1 Incubator set at 35°C±0.5°C with provision for maintaining materials at above 80% humidity.

6.2 Filter funnel apparatus for 47 mm membrane filters, with a vacuum source.

6.3 Dissecting microscope (10-15X) with built-in light sources.

6.4 Sterile disposable or properly cleaned (by well known standard methods) glass or plasticware including 1 and 10 mL pipettes, sample collection containers, flasks, graduated cylinders, and diluent containers.

6.5. Sterile forceps

6.6 Sterile 50 mm diameter petri dishes with absorbent pads

6.7 Sterile 0.45μm pore size, 47 mm diameter micropore filters for sample filtering

6.8 Sterile Dilution and Rinse water prepared in accordance with standard methods.

6.9 Biohazard bag

7.0 Reagents and Standards

7.1 Coliscan® MF medium of Micrology Laboratories LLC is provided in a broth for using 2 mL/dish, or it may be obtained in dehydrated agar based form. The media are provided without the cefsulodin antibiotic (for the elimination of some non-target organisms). The medium should be kept frozen (2-6°C) and has an expiration time of 6 months.

7.2 Preparation of the Medium for use: The broth medium should be thawed and freshly prepared cefsulodin solution should be added so that the cefsulodin concentration equals 5 μg/mL of medium. The prepared medium can be kept refrigerated for up to 2 weeks.
The agar medium should be mixed with reagent-grade water according to instructions and autoclaved for 15 min/15 lb pressure, tempered to below 80° C and then freshly prepared cefsulodin solution should be added (5μg/mL of medium). The medium should then be dispensed into 50 mm petri dishes (5 mL/dish). Store at 4°C for up to 2 weeks.

7.3 Preparation of the Cefsulodin solution: For Coliscan® MF broth, pure cefsulodin is provided in a sterile tube to which 10 mL of sterile reagent-grade water (provided) is added. Shake until the cefsulodin is completely dissolved and dispense 0.5 mL of this solution into each 20 mL bottle of Coliscan® MF broth.
For Coliscan® agar medium, pure cefsulodin is provided in a sterile tube to which 10 mL of sterile reagent-grade water (provided) is added. Shake until the cefsulodin is completely dissolved and dispense the 10 mL aliquot into 1 liter of hot liquid autoclaved (tempered) Coliscan® agar, mix well and dispense into 50 mm petri dishes (5 mL/dish). Cefsulodin solutions will be sterile if carefully prepared to avoid the introduction of contaminating microbes. The prepared media should be

affirmed as sterile by incubating a control plate of each lot and verifying no growth.

If operators prefer, they can filter sterilize their cefsulodin solutions to ensure sterility, but testing the prepared medium with a control is still recommended.

7.4 Make a 10% solution of sodium thiosulfate using reagent-grade water.

8.0 Sample collection, Dechlorination, Preservation, Shipment and Storage 8.1 Collect samples in a sterile, clean wide mouth glass or heat resistant plastic bottle with a leakproof closure, all of which is non-toxic in use. A presterilized, sealable, non-toxic plastic bag may also be used for sample collection.

8.2 For potable water, open the tap and allow the water to run for 2-3 minutes and then collect the sample using aseptic technique to avoid contamination. For other sample types, aseptically collect water that is representative of the source.

8.3 Samples with residual chlorine should be neutralized at the time of collection by adding 1 mL of a 10% solution of sodium thiosulfate (or the equivalent) per liter of sample.

8.4 Samples should be tested as soon as possible after collection. If processing is not done within 1 hour, the sample should be held on ice or refrigerated at 2-8°C. Potable water samples should be tested or processed within a maximum holding time of 30 hours of collection and non-potable samples should be tested or processed within a maximum holding time of 8 hours of collection.

9.0 Quality Control

9.1 Coliscan® MF is tested for quality control at the time of manufacture and is certified to meet specifications. Each lot should be tested by the using laboratory by preparing three plates of the medium, one to serve as a positive control, one to serve as a negative control, and one to serve as an uninoculated control.

Prepare 24 hour tryptone broth cultures of typical E. coli, Enterobacter aerogenes or Klebsiellsa pneumonia, and Pseudomonas aeruginosa or Salmonella typhimurium. Prepare serial dilutions of E. coli and Enterobacter or Klebsiella combined so that the combined inoculum will result in 20-80 CFU/100 mL and filter. Place the filter on the surface of one of the plates of Coliscan® MF medium. Prepare serial dilutions of the Pseudomonas or Salmonella so that the inoculum will result in 20-80 CFU/100 mL and filter. Place the filter on the surface of the second plate of Coliscan® MF medium. Filter 100 mL of sterile diluent and place the filter on the surface of the third plate of Coliscan® MF medium. Use sterile filter apparatus for each prep. Incubate the plates 24 hr. at 35°C±0.5°C. The E. coli/Enterobacter or Klebsiella control should have both blue/purple (E. coli) and pink/magenta (Enterobacter or Klebsiella) colonies, the Pseudomonas or Salmonella control should have colorless colonies, and the diluent blank control should have no colonies.

Colonies from the controls may be picked and tested further with various diagnostic media if desired.

10.0 Calibration and Standardization

10.1 Coliscan® MF calibration or standardization is not required.

10.2 Incubators should be tested daily to assure maintenance of proper temperature. Thermometers used should be tested at least annually against an NIST certified thermometer.

11.0 Procedure

11.1 Test Procedure

11.1.2 Using proper technique, filter the sample through a 47 mm, 0.45μm pore size membrane filter. Rinse the filter funnel twice with at least 20 mL of sterile diluent/rinse to complete the filtration. Transfer the filter to a petri dish containing a pad saturated with 2 mL of the Coliscan® MF broth, invert the dish and incubate at 35°C±0.5°C for 24 hours.

11.2 Interpretation

11.2.1 Check the filters for colony forming units. Generally, colonies are obvious and can be observed with the unaided eye in normal room or daylight. However, the use of a 10-15X magnifying device is recommended for critical analysis.

11.2.2 The sum of blue/purple and pink/magenta colonies is the Total Coliform Positive count. Blue/purple colonies are counted as E. coli. Pink/magenta colonies are counted as other than E. coli coliforms. Clear or white colonies are counted as non-coliforms.

12.0 Data Analysis, Calculation, Interpretation and Reporting Results

12.1 Presence/Absence Test

12.1.1 The presence of at least one blue/purple or pink/magenta colony at least 0.5 mm in diameter indicates the sample is total coliform positive. The presence of at least one blue/purple colony indicates the sample is positive for E. coli. The presence of at least one pink/magenta colony indicates the sample is positive for general coliforms.

12.2 General Coliform (excludes E. coli) - Quantitative Test

12.2.1 Count the number of pink/magenta CFUs present on the membrane filter and record as the number/amount of sample used for that test. For example, if the amount of sample was 10 mL and there were 20 pink CFUs, record as 20 per 10 mL. Then translate to the number of CFUs/100 mL. In this case, the 10 mL sample is 0.1 of 100, so the 20 CFUs should be multiplied X10, giving 200 CFU/100 mL sample. All pink/magenta CFUs should be counted as general coliforms. (Colonies should be at least 0.5 mm diameter to be counted.)

12.3 E. coli (Fecal) - Quantitative Test

12.3.1 Count the number of blue/purple CFUs present on the membrane filter and record as the number of E. coli/amount of sample used for that test. Then translate to the number of E. coli CFUs/100 mL of sample (see 12.1.1). All blue/purple CFUs should be counted as E. coli. (Colonies should be at least 0.5 mm diameter to be counted.)

12.4 Total Coliforms - Quantitative Test

12.4.1 The sum of the number of general coliform CFUs and the number of E. coli CFUs from one sample equals the number of total coliforms in that sample. That is, the total number of pink/magenta CFUs and blue/purple CFUs = the total coliforms for that sample.

13.0 Method Performance Characteristics

13.1 Specificity - In a study done to compare Coliscan®MF to the m-TEC Method for the detection and enumeration of E. coli from disinfected wastewater effluent, the false positive error was 3.8% and the false negative error was less than 1.0%. That is, of 105 CFUs judged to be E. coli (blue/purple) which were picked and subjected to Enterotube® analysis, only 4 were identified as other than E. coli. And of 131 CFUs judged to be coliforms other than E. coli (pink/magenta) which were picked and subjected to Enterotube analysis, only 1 was identified as other than a true coliform.

13.2 Comparability - The Pearson Coefficient for the parallel analyses on the Coliscan®MF and the m-TEC methods within the same laboratory was 0.928. T- test analyses indicated no significant differences between the methods at the 95% confidence level.

13.3 The membrane filter medium known as MI Agar was developed by the USEPA and was approved for use with drinking water as stated and published in the Federal Register Vol. 64, No. 230, Dec. 1, 1999. On the basis that the Coliscan MF® medium is based upon very similar, equally effective technology, the EPA has granted it official approval for the determination of total coliforms and E. coli for use in drinking water monitoring also. The EPA reported the false-positive and false- negative rates for E. coli to both be 4.3% with the MI Agar method. The specificities for E. coli and total coliforms were reported to be 95.7% and 93.1%, respectively. Because of the parallels in the methodology between the MI Agar and the Coliscan®MF, testing results should be the same for both media.

14.0 Pollution Prevention

14.1 Laboratory personnel should use pollution control techniques to minimize waste generation wherever possible. Where this is not possible at the source, recycling should be practiced.

15.0 Waste Management

15.1 Each laboratory is responsible to comply with all federal, state and local regulations governing waste management. Special emphasis should be placed on hazardous waste identification rules and land disposal restrictions and to protecting the air, water, and land by minimizing and controlling all release from fume hoods and bench operations. Compliance is also required with any sewerage discharge permits and regulations. Federal, state or local authorities should be contacted for further specific information.

16.0 References

16.1 APHA (1995) Standard Methods for the Examination of Water and Wastewater. Edition 19.

16.2 Brenner, K.P., et al. (2000) Membrane Filter Agar Medium Containing Two Enzyme Substrates Used for the Simultaneous Detection of Total Coliforms and E. coli. United States Patent #6,063,590.

16.3 Brenner, K.P., et al. (1993) New Medium for the Simultaneous Detection of Total Coliforms and Escherichia coli in Water. Appl. Environ. Microbiol. 59: 3534-3544.

16.4 Roth, J.N., W.J. Ferguson. (1993) Method Test Media and Chromogenic Compounds for Identifying and Differentiating General Coliforms and Escherichia coli Bacteria. United States Patent #5,210,022.

16.5 Umble, A.K., et al. (1999) Elkhart, Ind., Tests an Improved, Simplified Membrane Filtration Method for Escherichia coli Detection and Enumeration. Water Environment Tech. Vol.11, No.4, 57-59.

16.6 USEPA. National primary and secondary drinking water regulations: Analytical methods for regulated drinking water contaminants: Proposed rule, F6. Federal register 58(129): 65626. Wash., D.C., Office of the Federal Register. Dec. 15, 1993.

16.7 USEPA. National primary drinking water regulations: Analytical methods for certain pesticides and microbial contaminants: 40 CFR Part 141. Federal register Vol. 63, No. 147, July 31, 1998. Proposed rules.

Appendix 11

Chlorine Dioxide
http://home.windstream.net/mikeric/Odor/clo2.htm

What You Can't Tell From the Name

While chlorine dioxide has chlorine in its name, its chemistry is radically different from that of chlorine.

The way it works is almost magical. It has to do with the way electrons interact with one another. As we all learned in high school chemistry, we can mix two compounds and create a third that bears little resemblance to its parents.

For instance: Mix two parts of hydrogen gas with one of oxygen and liquid water is the result.

Mix equal parts of caustic soda (commonly called lye, a part of everyday soap) and hydrochloric acid (which will dissolve iron) and you get table salt and water. And for chlorine dioxide, mix one part chlorine gas with two parts of oxygen.

We should not be misled by the fact that chlorine and chlorine dioxide share a word in common.

Hydrogen is in both water and in hydrogen cyanide. The latter can be a deadly poison.

At room temperature, chlorine is a greenish-yellow poisonous gas. When added to water, however, chlorine combines with water to form hypochlorous acid that then ionizes to form hypochlorite ion - 'bleach'

Regarding bleaching, chlorine dioxide and chlorine -- because of their fundamentally different chemistries -- react in distinct ways with organic compounds, and as a result generate very different byproducts.

It is this difference that explains the superior environmental performance of chlorine dioxide in paper making and scrubbers.

Technically speaking, both chlorine and chlorine dioxide are oxidizing agents -- electron receivers.

Chlorine has the capacity to take in two electrons, whereas chlorine dioxide can absorb five.

This property, along with the complex, but well known, ways chlorine combines with lignin (the cellular adhesive in wood tissue), explains the basic difference between the two compounds.

In the chlorine-based bleaching process, about 10 percent of the chlorine combines directly with lignin which has "aromatic" components.

Aromatic compounds have atoms arranged in rings, and they may have other atoms, such as chlorine, attached to these rings.

Within the group of chlorinated aromatics, which can be toxic to some organisms, are the infamous dioxins.

Chlorine dioxide's behavior as a bleaching agent is quite dissimilar. Instead of combining with the aromatic rings, chlorine dioxide breaks these rings apart.

In addition, as the use of chlorine dioxide increases, the generation of chlorinated organics falls dramatically.

Chlorine dioxide's chemistry explains why it is such an effective oxidant, or bleaching agent.

It is 2.5 times more active than chlorine gas, and much more selective.

Chlorine dioxide attacks the lignin, but does not react with the desired cellulose in wood tissue. It is cellulose -- the tree's fiber -- that provides the strength in the final paper products.

These advantages make chlorine dioxide the preferred environmental standard for eliminating toxic substances in mill waste water and scrubbers.

Chlorine dioxide is a neutral compound of chlorine in the +IV oxidation state.

It has a boiling point of 11 degrees C at atmospheric pressure.

The liquid is denser than water, and the gas is denser than air.

The molecule is polar with the oxygen atoms separated by 116.5 degrees.

Water (H_2O) is also polar (105 degrees).

The presence of organic matter, nutrients, and microorganisms in the output of sewage treatment plants is measured by three tests: coliform count, algal count, and biochemical oxygen demand (BOD).

The coliform count describes the number of E. coli (the characteristic bacteria in animal wastes) present.

The algal count is a biological test for microorganisms other than bacteria and viruses which may be present.

The BOD measures the volume of oxygen gas taken up by a given amount of water in five days at 20 degrees C, (remember, there is an ultimate test of BOD).

The biochemical oxygen demand analysis is an attempt to simulate the effect a waste will have on the dissolved oxygen of a stream by a laboratory test.

The BOD test gives an indication of the amount of oxygen needed to stabilize or biologically oxidize the waste.

The advantage of the BOD test is that it measures only the organics which are oxidized by the bacteria.

The disadvantage is the 5 day time lag and the difficulty in obtaining consistent repetitive values.

The results of the COD (chemical oxygen demand) tests are usually higher that the corresponding BOD test for the following reasons:

Many organic compounds which are dichromate oxidizable are not biochemically oxidizable.

Certain inorganic substances, such as sulfides, sulfites, thiosulfates, nitrites and ferrous iron are oxidized by dichromate, creating an inorganic COD, which is misleading when estimating the organic content of the wastewater.

The BOD results may be affected by lack of seed acclimation, giving erroneously low readings. The COD results are independent of seed acclimation.

In general, chlorine dioxide has been found to produce fewer organic byproducts with naturally occurring dissolved organic material.

Chlorine dioxide is an explosive gas, but is stable in water in the absence of light and elevated temperatures ... which is just what we do.

ClO_2 is capable of oxidizing iron and manganese, removing color, and lowering THM (Trihalomethanes) formation potential.

It also oxidizes many organic and sulfurous compounds that cause off-tastes and odors.

Chlorine dioxide is a green-yellow gas that decomposes readily and with explosive force to chlorine and oxygen. It is, therefore, usually manufactured on-site.

Chlorine dioxide is a more powerful biocide than free chlorine but does not persist as long as chlorine

Appendix 12

Barium and Compounds (CASRN 7440-39-3)
http://www.epa.gov/iris/subst/0010.htm

I. Chronic Health Hazard Assessments for Noncarcinogenic Effects

I.A. Reference Dose for Chronic Oral Exposure (RfD)

Barium and Compounds
CASRN — 7440-39-3
Section I.A. Last Revised — 07/05/2005

The RfD is an estimate of an exposure, designated by duration and route, to the human population (including susceptible subgroups) that is likely to be without an appreciable risk of adverse health effects over a lifetime. It is derived from a statistical lower confidence limit on the benchmark dose (BMDL), a no-observed-adverse-effect level (NOAEL), a lowest-observed-adverse-effect level (LOAEL), or another suitable point of departure, with uncertainty/variability factors applied to reflect limitations of the data used. The RfD is intended for use in risk assessments for health effects known or assumed to be produced through a nonlinear (possibly threshold) mode of action. It is expressed in units of mg/kg-day. Please refer to the guidance documents at http://www.epa.gov/iris/backgrd.html for an elaboration of these concepts. Since RfDs can be derived for the noncarcinogenic health effects of substances that are also carcinogens, it is essential to refer to other sources of information concerning the carcinogenicity of this chemical substance. If the U.S. EPA has evaluated this substance for potential human carcinogenicity, a summary of that evaluation will be contained in Section II of this file.

An RfD of 7×10-2 mg/kg-day was previously entered on the IRIS data base in 1998. This value was based on a NOAEL of 0.21 mg/kg-day for the absence of a hypertensive effect in two human studies (Brenniman and Levy, 1984; Wones et al. 1990). The subchronic and chronic rat NTP (1994) studies, and the McCauley et al. (1985) study of unilaterally nephrectomized rats were used to support the identification of the kidney as a co-critical target. An uncertainty factor of 3 was used to account for data base deficiencies.

The change in the value of the RfD from the previous IRIS assessment is due to selection of a new principal study and critical effect, the use of benchmark dose modeling to determine the point of departure, and a new evaluation of both the literature and application of uncertainty factors.

I.A.2. Principal and Supporting Studies

NTP (1994) exposed both sexes of F344/N rats and both sexes of B6C3F1 mice to barium chloride dihydrate (BaCl2×2H2O) in drinking water for 13 weeks or 2 years. Drinking water concentrations in the chronic study (60 animals/sex/group) were and 0, 500, 1250, and 2500 ppm. The study authors estimated doses, using water consumption and body weight data, as 0, 15, 30, and 60 mg Ba/kg-day for male rats and 0, 15, 45, and 75 mg Ba/kg-day for female rats. The estimated doses for mice were 30, 75, and 160 mg Ba/kg-day for males and 40, 90, and 200 mg Ba/kg-day for females. In the subchronic study (10 animals/sex/group), drinking water concentrations were 0, 125, 500, 1000, 2000, and 4000 ppm. Estimated doses were 0, 10, 30, 65, 110, and 200 mg Ba/kg-day for male rats and 0, 10, 35, 65, 115, and 180 mg Ba/kg-day for female rats. For mice, the corresponding estimated doses were 0, 15, 55, 100, 205, and 450 mg Ba/kg-day for the males and 0, 15, 60, 110, 200, and 495 mg Ba/kg-day for the females. The animals were fed an NIH-07 diet. Barium was not reported as a contaminant of the feed.

Chemical-related nephropathy was observed in male and female mice following chronic or subchronic drinking water exposure to barium chloride. These lesions were characterized by tubule dilatation, renal tubule atrophy, tubule cell regeneration, hyaline cast formation, multifocal interstitial fibrosis, and the presence of crystals, primarily in the lumen of the renal tubules. NTP pathologists concluded that these lesions were morphologically distinct from the spontaneous degenerative renal lesions commonly observed in aging mice. Survival rates were significantly reduced in the high dose group by 65% for males and 26% females when compared to controls. Mortalities were attributed to the chemical-related renal lesions (NTP, 1994). A significant number of chronically exposed mice in the high dose group had mild to severe cases of nephropathy, 19/60 males and 37/60 females. One female and two males in the intermediate dose group had mild to moderate cases of chemical-related nephropathy. Chemical-related nephropathy was also observed in rats following subchronic exposure. In the chronic rat study, spontaneous nephropathy was observed in the majority of animals in both control and treatment groups precluding the detection of any treatment-related effect. Increased kidney weights were observed in male and female rats and female mice following 13 weeks of exposure. Female rats were the only animals with increased kidney weights following 15 months of exposure. See Section 4 of the Toxicological Review (U.S. EPA, 2005) for additional details.

The kidney appears to be the most sensitive target of toxicity resulting from repeated ingestion of soluble barium salts. Chronic and subchronic rodent studies conducted by NTP (1994) and McCauley et al. (1985) provide evidence for an association between barium exposure and renal toxicity. However, chronic and subchronic rodent studies conducted by Tardiff et al., (1980) and Schroeder and Mitchener (1975a, b) were unable to detect adverse effects, including renal toxicity, following exposure to barium. Unfortunately, no human studies have investigated the effects of barium exposure on the kidneys. The NTP (1994) 2-year drinking water study in B6C3F1 mice was selected as the principal study and chemical-related nephropathy was identified as the critical effect for deriving an RfD for barium and its soluble salts. The principal study and critical effect were selected after careful evaluation of all the available toxicity studies. The primary reason for selecting this study and critical effect was that the nephropathy data provide the best evidence of a dose-response relationship.

There is conflicting evidence whether or not barium exposure may induce hypertensive effects. An investigation of anesthetized dogs (Roza and Berman, 1971) infused with barium chloride at a rate of 2 μmol/kg/min reported an increase in mean blood pressure from 138/86 to 204/103 (n =24). In a series of subchronic and chronic drinking water studies, Perry et al. (1989, 1985) observed a hypertensive effect in rats receiving as little as 6 mg/kg-day. The animals in these studies were maintained on a low metal diet with lower concentrations of calcium and other minerals than standard rat chow. However, NTP (1994) found no association between subchronic barium exposure and cardiovascular toxicity in rats at the highest level tested (200 mg/kg-day). Likewise, McCauley et al. (1985) observed no adverse effect on blood pressure following subchronic exposure to barium in drinking water at the highest level tested (150 mg/kg-day).

The reduced concentrations of calcium and other minerals in the low metal diet has been identified as a possible reason for the discrepancy between the findings of Perry et al. (1989, 1985) and other animal studies that did not observe hypertension in barium-treated animals (NTP, 1994; McCauley et al.,1985). The calcium concentration in the low metal diet was 3.8 g/kg, while the nutritional requirement for maintenance, growth, and reproduction of rats is 5 g/kg (NRC, 1995). Perry has stated that the concentration of calcium in the diet was adequate for normal growth and development (Perry, 1984). It is, however, unclear if the reduced dietary concentrations of calcium may have contributed to development of barium-related hypertension. There is some evidence that reduced dietary calcium is a risk factor for hypertension in humans (McCarron et al., 1984). In light of the possible association between reduced calcium intake and hypertension, and because hypertension has not been reported in animals receiving the recommended dietary concentration of calcium, the data from Perry et al. (1989, 1985) were not considered in the derivation of the RfD.

Acute hypertension has been observed in humans following accidental or intentional ingestion of soluble barium salts (CDC, 2003; Downs et al., 1995). Two human studies have investigated the effects of longer-term barium ingestion on blood pressure (Brenniman et al., 1981; Wones et al.,1990). Both investigations found no hypertensive effect with their highest exposure concentrations. Brenniman et al. (1981) found no effect on hypertension between two communities with a 70-fold difference in the barium concentrations of their drinking water. Wones et al. (1990) found no hypertensive effect in a before-and-after comparison of 11 subjects that were exposed to two concentrations of barium in their drinking water over the course of ten weeks. Coincidently, the same NOAEL of 0.21 mg/kg-day was identified for both studies. These NOAELs were estimated by EPA using standard estimates for drinking water intake (2 L/day) and average body weight (70 kg).

Neither Brenniman et al. (1981) nor Wones et al. (1990) provided sufficient data to support, or refute, the hypothesis that chronic barium exposure causes hypertension. Hypertension is a complex multifactorial condition and it is very possible that the effect of chronic barium exposure on blood pressure is relatively small compared to other determinates such as diet and exercise. Wones et al. (1990) attempted to control for the effect of diet by providing a standard diet to all of the study participants. Unfortunately, the power of this study was limited by the very small number of participants (n=11). They also used short exposure durations (4 weeks for each exposure concentration), which may not have been sufficient to observe a chronic effect. Brenniman et al. (1981) also examined a relatively small number of subjects (n=85) in the subpopulation that was controlled for key risk factors. Other limitations of Brenniman et al. (1981) were that they collected replicate blood pressure measurements from individuals during a single 20-minute period, they used community-wide exposure estimates, and they didn't control for a number of important risk factors for hypertension, including diet and exercise. In the absence of dose-response data for barium-induced hypertension the RfD was not based this effect.

The RfD was derived by the benchmark dose approach using renal lesions in mice as the critical effect from the NTP (1994) study. The benchmark response predicted to affect 5% of the population (BMR05) was selected for the point of departure. Benchmark analysis (BMDS version 1.3.2) was used to model the incidence of chemical-related nephropathy in male and female mice (NTP, 1994). The BMD05 for males was 84 mg/kg-day and the lower 95% confidence limit (i.e., BMDL05) was 63 mg/kg-day. The BMD05 for females was 93 mg/kg-day and the BMDL05 was 58 mg/kg-day. These BMDL05 values are very similar, but since there is slightly less uncertainty in the estimate derived from the male mice (the BMD05 and BMDL05 are closer together), the male BMDL05 was used for deriving the RfD.

A lower BMR than 10% extra risk could be used because the critical effect was considered to be substantially adverse and distinctly chemical-related, and because the data range included a response lower than 10% (U.S. EPA, 2002). First, the lesions in the intermediate dose group (severity grades, mild to moderate) were intermediate on a continuum leading to severe nephropathy, with severity between that seen in the control group (maximum severity grade, minimal) and the high dose group (severity grades, mild to marked). Since the significantly reduced survival rate in the high dose group was associated with the chemical-related renal lesions (NTP, 1994), the effects in the intermediate dose group are considered possibly irreversible, and biologically significant. Further, a similar pattern of effects was evident in both males and females.

I.A.3. Uncertainty and Modifying Factors (Oral RfD)

UF = 300

A total UF of 300 was applied: 10 for extrapolation for interspecies differences (UFA: animal to human); 10 for consideration of intraspecies variation (UFH: human variability); and 3 for deficiencies in the database (UFD). A value of 10 for both the interspecies and intraspecies UFs is generally used in the absence of data to indicate otherwise. The rationale for application of the UFs is described below.

A 10-fold UF was used to account for uncertainty in extrapolating from laboratory animals to humans (i.e., interspecies variability). Insufficient information is available regarding the toxicity of chronic barium exposure to compare the dose-response relationship in animals with what could be expected in humans. No information was available to quantitatively assess toxicokinetic or toxicodynamic differences between animals and humans.

A 10-fold UF was used to account for variation in susceptibility among members of the human population (i.e., interindividual variability). This UF was not reduced from a default of 10 because there are insufficient data on the dose-response relationship in humans and because there are studies in experimental animals that suggest gastrointestinal absorption may be higher in children than in adults (Taylor et al., 1962; Cuddihy and Griffith, 1972).

A 3-fold UF was used to account for uncertainty associated with deficiencies in the data base. The database of oral barium toxicity consists of two human studies, which found no effect on hypertension (Brenniman et al., 1981; Wones et al., 1990), and several chronic and subchronic rodent studies. The database is deficient in several areas: neither a two-generation reproductive toxicity study nor an adequate investigation of developmental toxicity has been conducted. It is also not known if barium deposition in bone tissue is associated with an adverse effect. The available data indicate that renal toxicity is likely to be the most sensitive endpoint for chronic barium exposure.

An UF was not needed to account for subchronic- to-chronic extrapolation because a chronic study was used to derive the RfD. An UF for LOAEL-to-NOAEL extrapolation was not used since benchmark dose modeling was employed to determine the point of departure.

I.A.4. Additional Studies/Comments (Oral RfD)

Several studies have investigated the hypertensive effect of barium in animals. Intravenous infusion of barium chloride into anesthetized dogs or guinea pigs resulted in increased blood pressure and cardiac arrhythmias (Hicks et al., 1986; Roza and Berman, 1971). Perry et al. (1989, 1985, 1983) were the only authors to report hypertension in animals following long-term ingestion of barium. The rats in this study were maintained on a rye-based diet with a calcium content below the recommended daily requirement (NRC, 1995). The diet was also lower in potassium than standard rat chow. Animals maintained on diets low in calcium or potassium may be more sensitive to the cardiovascular effects of barium. Acute effects of barium on the cardiovascular system have been shown to be modified by calcium and potassium (Shanbaky et al., 1978; Roza and Berman, 1971). Barium has also been shown to be a calcium agonist (U.S. EPA, 1990; WHO, 1990; Perry et al., 1989; Brenniman et al., 1981; Shanbaky et al., 1978). Potassium alleviates the cardiac effects and skeletal muscle effects associated with acute barium poisoning (U.S. EPA, 1990; WHO, 1990; Gould et al., 1973; Roza and Berman, 1971; Diengott et al., 1964). Perry and Erlanger (1982) observed that rats maintained on the rye-based diet and exposed to cadmium developed hypertension, whereas rats maintained on standard chow and exposed to cadmium did not. In view of a possible association between the barium-induced cardiovascular effects and calcium and potassium intake, the relevance of the data from Perry et al. (1983) to animals maintained on standard diets, or humans is uncertain.

NTP (1994) evaluated blood pressure and electrocardiogram readings of rats exposed to barium in drinking water for 13 weeks. No association was detected between subchronic barium exposure and cardiovascular toxicity in rats at the highest level tested (200 mg/kg-day). Likewise, McCauley et al. (1985) observed no adverse effect on blood pressure following administration of barium in drinking water at the highest level tested (150 mg/kg-day).

The uptake of barium in bone tissue was evaluated in F344/N rats sacrificed at the 15-month interim of the NTP (1994) 2-year drinking water study. Barium concentrations in upper, middle, and lower sections of the femur were increased by approximately three-orders of magnitude in the high dose groups when compared to controls. Minimal reductions in calcium concentrations were observed in the same femur sections and no effect on bone density was observed. The biological implications of increased barium deposition in the bone tissue is unclear. It is possible that barium may interfere with the physiological processes of bone tissue including white blood cell production. A significant reduction in mononuclear cell leukemia was observed in treated male rats (NTP, 1994). Addition research is needed to fully investigate potential osteogenic effects of elevated barium exposure.

Dietz et al. (1992) evaluated the reproductive toxicity of barium in rats and mice. A dose of approximately 200 mg/kg-day was associated with lower pregnancy rates in rats and mice. However, below normal pregnancy rates were observed in rats from both the treatment and control groups. A non-significant reduction in litter size was observed in rats that received an approximate dose of 200 mg/kg-day. In mice there was a significant reduction in litter size in animals receiving a dose of approximately 100 mg/kg-day, but the effect was not observed in the higher dose group. Birth weight in rat pups was significantly reduced in the 200 mg/kg-day treatment group, but no effect was observed on postnatal day 5. Based on this limited data set it is not clear if barium is associated with reproductive toxicity.

VI.A. Oral RfD References

Brenniman, GR; Levy, PS. (1984) Epidemiological study of barium in Illinois drinking water supplies. In: Advances in modern toxicology. Calabrese, EJ, ed. Princeton, NJ: Princeton Scientific Publications, pp. 231-249.

Brenniman, GR; Kojola, WH; Levy, PS; et al. (1981) High barium levels in public drinking water and its association with elevated blood pressure. Arch Environ Health 36(1):28-32.

Centers for Disease Control (CDC). (2003) Barium toxicity after exposure to contaminated contrast solution - Goias State, Brazil, 2003. Available on-line at: http://www.cdc.gov/mmwr/preview/mmwrhtml/mm5243a5.htm. Accessed 03/10/2004.

Cuddihy, RG; Griffith, WC. (1972) A biological model describing tissue distribution and whole-body retention of barium and lanthanum in beagle dogs after inhalation and gavage. Health Phys 23:621-633.

Diengott, D; Rozsa, O; Levy, N; et al. (1964) Hypokalaemia in barium poisoning. Lancet 14:343-344.

Dietz, DD; Elwell, MR; Davis, WE, Jr.; et al. (1992) Subchronic toxicity of barium chloride dihydrate administered to rats and mice in the drinking water. Fund Appl Toxicol 19:527-537.

Downs, JC; Milling D; Nichols, CA. (1995) Suicidal ingestion of barium-sulfate-containing shaving powder. Am J Forensic Med Pathol 16:56-61.

Gould, DB; Sorrell, MR; Lupariello, AD. (1973) Barium sulfide poisoning. Arch Intern Med 132:891-894.

Hicks, R; Caldas, LQ; Dare, PRM; et al. (1986) Cardiotoxic and bronchoconstrictor effects of industrial metal fumes containing barium. Arch Toxicol Suppl 9:416-420.

McCarron, DA; Morris, CD; Henry, HJ; et al. (1984) Blood pressure and nutrient intake in the United States. Science 224:1392-1398.

McCauley, PT; Douglas, BH; Laurie, RD; et al. (1985) Investigations into the effect of drinking water barium on rats. In: Inorganics in drinking water and cardiovascular disease. Calabrese, EJ, ed. Princeton, NJ: Princeton Scientific Publications, pp. 197-210.

National Research Council (NRC). (1995) Nutrient requirements of laboratory animals. Washington, DC: National Academy Press, p. 13.

National Toxicology Program (NTP), Public Health Service, U.S. Department of Health and Human Services. (1994) NTP technical report on the toxicology and carcinogenesis studies of barium chloride dihydrate (CAS no. 10326-27-9) in F344/N rats and B6C3F1 mice (drinking water studies). NTP TR 432. Research Triangle Park, NC. NIH pub. no. 94-3163. NTIS pub PB94-214178.

NTP (2004) Nonneoplastic Lesions by Individual Animal - Barium Chloride Dihydrate. Available on-line: http://ntp-server.niehs.nih.gov/index.cfm?objectid=037BBD0D-F9EB-7773-1E4ECB464EC0DF30. Accessed 05/10/04.

Perry, HM. (1984) Discussion in: Advances in Modern Toxicology. Calabrese, EJ, ed. Princeton, NJ: Princeton Scientific Publications, pp. 241-249.

Perry, HM, Jr.; Erlanger, MW. (1982) Effect of diet on increases in systolic pressure induced in rats by chronic cadmium feeding. J Nutr 112:1983-1989.

Perry, HM, Jr; Kopp, SJ; Erlanger, MW; et al. (1983) Cardiovascular effects of chronic barium ingestion. In: Hemphill, DD, ed. Trace substances in environmental health. XVII, Proceedings of University of Missouri's 17th Annual Conference on Trace Substances in Environmental Health. Columbia, MO: University of Missouri Press, pp. 155-164.

Perry, HM, Jr.; Perry, EF; Erlanger, MW; et al. (1985) Barium-induced hypertension. Ch. XX. Adv Mod Environ Toxicol 9:221-279.

Perry, HM, Jr.; Kopp, SJ; Perry, EF; et al. (1989) Hypertension and associated cardiovascular abnormalities induced by chronic barium feeding. J Toxicol Environ Health 28(3):373-388.

Roza, O; Berman, LB. (1971) The pathophysiology of barium: hypokalemic and cardiovascular effects. J Pharmacol Exp Ther 177:433-439.

Schroeder, H; Mitchener, M. (1975a) Life-term studies in rats: effects of aluminum, barium, beryllium and tungsten. J Nutr 105:421-427.

Schroeder, H; Mitchener, M. (1975b) Life-term effects of mercury, methyl mercury and nine other trace metals on mice. J Nutr 105:452-458.

Shanbaky, IO; Borowitz, JL; Kessler, WV. (1978) Mechanisms of cadmium- and barium-induced adrenal catecholamine release. Toxicol Appl Pharmacol 44:99-105.

Tardiff, RG; Robinson, M; Ulmer, NS. (1980) Subchronic oral toxicity of BaCl2 in rats. J Environ Pathol Toxicol 4:267-275.

Taylor, DM; Pligh, PH; Duggan, MH. (1962) The absorption of calcium, strontium, barium and radium from the gastrointestinal tract of the rat. Biochem J 83:25-29.

U.S. Environmental Protection Agency (U.S. EPA). (1990) Drinking water criteria document on barium. Prepared by the Office of Health and Environmental Assessment, Cincinnati, OH, for the Criteria and Standards Division, Office of Drinking Water, Washington, DC, EPA/NTIS PB91-142869.

U.S. EPA (2000) Benchmark dose technical guidance document [external review draft]. EPA/630/R-00/001. Available from: http://www.epa.gov/iris/backgrd.html.

U.S. EPA (2002) A review of the reference dose and reference concentration processes. Risk Assessment Forum, Washington, DC; EPA/630/P-02/0002F. Available from: http://www.epa.gov/iris/backgrd.html.

U.S. EPA (2005) Toxicological review of barium and compounds. Integrated Risk Information System (IRIS). National Center for Environmental Assessment, Washington, DC; NCEA-S-1683. Available from: http://www.epa.gov/iris.

Wones, RG; Stadler, BL; Frohman, LA. (1990) Lack of effect of drinking water barium on cardiovascular risk factor. Environ Health Perspect 85:355-359.World Health Organization (WHO). (1990) Environmental health criteria 107: barium. Sponsored by United Nations Environment Programme, International Labour Organisation, and World Health Organization. Geneva, Switzerland.

Appendix 13

Boron and Compounds (CASRN 7440-42-8)
http://www.epa.gov/iris/subst/0410.htm

Boron and Compounds; CASRN 7440-42-8 (08/05/2004)
I. Chronic Health Hazard Assessments for Noncarcinogenic Effects
I.A. Reference Dose for Chronic Oral Exposure (RfD)

In general, the oral Reference Dose (RfD) is an estimate (with uncertainty spanning perhaps an order of magnitude) of a daily exposure to the human population (including sensitive subgroups) that is likely to be without an appreciable risk of deleterious effects during a lifetime. The RfD is based on the assumption that thresholds exist for certain toxic effects such as cellular necrosis and is expressed in units of mg/kg-day. Please refer to the guidance documents at http://www.epa.gov/iris/backgrd.html for an elaboration of these concepts. Since RfDs can be derived for the noncarcinogenic health effects of substances that are also carcinogens, it is essential to refer to other sources of information concerning the carcinogenicity of this chemical substance. If the U.S. EPA has evaluated this substance for potential human carcinogenicity, a summary of that evaluation will be contained in Section II of this file.

This RfD replaces the previous RfD of 0.09 mg/kg-day entered on IRIS on 10/01/89 (see section VII. Revision History). Chronic toxicity in dogs (Weir and Fisher, 1972) was used previously to develop the RfD for boron. Recently, developmental data in three species (rats, mice, and rabbits) have become available. Based on the new developmental data and several limitations of the dog studies (Section I.A.1), decreased fetal body weight in rats is recommended as the critical effect for development of an RfD.

I.A.1. Oral RfD Summary

The BMDL05 was derived by Allen et al. (1996) using combined data from Price et al. (1996a) and Heindel et al. (1992). The BMR of a 5% decrease in fetal weight, relative to control, was selected for several reasons to help identify the point of departure. The dose response data (Price et al., 1996a) showed a statistically significant trend of decreasing fetal weights with increasing exposure to boron throughout the range of exposures tested. The exposure associated with the 5% weight decrease fell well within the range of the experimental data. Although the responses at lower doses were also lower than control response, the data base for boron is mixed concerning whether decreased fetal weights indicate a transient or more permanent functional alteration. For example, decreased weights did not persist in the companion study (Phase II of Price et al., 1996a, 1994). Therefore, no further adjustments were considered for identifying a level of oral exposure to boron associated with minimal level of risk.

__I.A.2. Principal and Supporting Studies (Oral RfD)

Heindel, JJ; Price, CJ; Field, EA; et al. (1992) Developmental toxicity of boric acid in mice and rats. Fund Appl Toxicol 18:266-277.

Price, CJ; Strong, PL; Marr, MC; Myers, CB; Murray, FJ. (1996a.) Developmental toxicity NOAEL and postnatal recovery in rats fed boric acid during gestation. Fund Appl Toxicol 32:179-193.

Developmental (decreased fetal weights) effects are considered the critical effect. The basis for calculating the RfD is the BMDL05 of 10.3 mg boron/kg-day calculated from the developmental effects reported by Heindel et al. (1992) and Price et al. (1996a).

Heindel et al. (1992) and Price et al. (1990) treated timed-mated Sprague-Dawley rats (29/group) with a diet containing 0, 0.1, 0.2, or 0.4% boric acid from gestation day (gd) 0-20. The investigators estimated that the diet provided 0, 78, 163, or 330 mg boric acid/kg-day (0, 13.6, 28.5 or 57.7 mg B/kg-day). Additional groups of 14 rats each received boric acid at 0 or 0.8% in the diet (539 mg/kg-day or 94.2 mg B/kg-day) on gd 6-15 only. Exposure to 0.8% was limited to the period of major organogenesis in order to reduce the preimplantation loss and early embryolethality indicated by the range-finding study and, hence, provide more opportunity for teratogenesis. (The range-finding study found that exposure to 0.8% on gd 0-20 resulted in a decreased pregnancy rate [75% as compared with 87.5% in controls] and in greatly increased resorption rate per litter [76% as compared with 7% in controls]). Food and water intake, and body weights, as well as clinical signs of toxicity, were monitored throughout pregnancy. On gd 20, the animals were sacrificed and the liver, kidneys, and intact uteri were weighed, and corpora lutea were counted. Maternal kidneys, selected randomly (10 dams/group), were processed for microscopic evaluation. Live fetuses were dissected from the uterus, weighed, and examined for external, visceral, and skeletal malformations. Statistical significance was established at p<0.05. There was no maternal mortality during treatment. Food intake increased 5-7% relative to that of controls on gd 12-20 at 0.2 and 0.4%; water intake was not significantly altered by administration of boric acid (data not shown). At 0.8%, water and food intake decreased on days 6-9 and increased on days 15-18, relative to controls. Pregnancy rates ranged between 90 and 100% for all groups of rats and appeared unrelated to treatment. Maternal effects attributed to treatment included a significant and dose-related increase in relative liver and kidney weights at 0.2% or more, a significant increase in absolute kidney weight at 0.8%, and a significant decrease in body-weight gain during treatment at 0.4% or more. Corrected body weight gain (gestational weight gain minus gravid uterine weight) was unaffected except for a significant increase at 0.4%. Examination of maternal kidney sections revealed minimal nephropathy in a few rats (unspecified number), but neither the incidence nor the severity of the changes was dose related.

Treatment with 0.8% boric acid (gd 6-15) significantly increased prenatal mortality; this was due to increases in the percentage of resorptions per litter and percentage of late fetal deaths per litter. The number of live fetuses per litter was also significantly decreased at 0.8%. Average fetal body weight (all fetuses or male or female fetuses) per litter was significantly reduced in all treated groups versus controls in a dose-related manner. Mean fetal weights were 94, 87, 63, and 46% of the corresponding control means for the 0.1, 0.2, 0.4 and 0.8% dose groups, respectively. The percentage of malformed fetuses per litter and the percentage of litters with at least one malformed fetus were significantly increased at 0.2% or more. Treatment with 0.2% or more boric acid also increased the incidence of litters with one or more fetuses with a skeletal malformation. The incidence of litters with one or more pups with a visceral or gross malformation was increased at 0.4 and 0.8%, respectively. The malformations consisted primarily of anomalies of the eyes, the central nervous system, the cardiovascular system, and the axial skeleton. In the 0.4 and 0.8% groups, the most common malformations were enlarged lateral ventricles of the brain and agenesis or shortening of rib XIII. The percentage of fetuses with variations per litter was reduced relative to controls in the 0.1 and 0.2% dosage groups (due primarily to a reduction in the incidence of rudimentary or full ribs at lumbar I), but was significantly increased in the 0.8% group. The variation with the highest incidence among fetuses was wavy ribs. Based on the changes in organ weights, a maternal lowest-observed-adverse-effect level (LOAEL) of 0.2% boric acid in the feed (28.5 mg B/kg-day) can be established; the maternal no-observed-adverse-effect level (NOAEL) is 0.1% or 13.6 mg B/kg-day. Based on the decrease in fetal body weight per litter, the level of 0.1% boric acid in the feed (13.6 mg B/kg-day) is a LOAEL; a NOAEL was not defined.

In a follow-up study, Price et al. (1996a, 1994) administered boric acid in the diet (at 0, 0.025, 0.050, 0.075, 0.100, or 0.200%) to timed-mated

CD rats, 60 per group, from gd 0-20. Throughout gestation, rats were monitored for body weight, clinical condition, and food and water intake. This experiment was conducted in two phases, and in both phases offspring were evaluated for post-implantation mortality, body weight and morphology (external, visceral, and skeletal). Phase I of this experiment was considered the teratology evaluation and was terminated on gd 20 when uterine contents were evaluated. The calculated average dose of boric acid consumed for Phase I dams was 19, 36, 55, 76, and 143 mg/kg-day (3.3, 6.3, 9.6, 13.3, and 25 mg B/kg-day). During Phase I, no maternal deaths occurred and no clinical symptoms were associated with boric acid exposure. Maternal body weights did not differ among groups during gestation, but statistically significant trend tests associated with decreased maternal body weight (gd 19 and 20 at sacrifice) and decreased maternal body weight gain (gd 15-18 and gd 0-20) were indicated. In the high-dose group, there was a 10% reduction (statistically significant in the trend test $p < 0.05$) in gravid uterine weight when compared with controls. The authors indicated that the decreasing trend of maternal body weight and weight gain during late gestation reflected reduced gravid uterine weight. Corrected maternal weight gain (maternal gestational weight gain minus gravid uterine weight) was not affected. Maternal food intake was only minimally affected at the highest dose and only during the first 3 days of dosing. Water intake was higher in the exposed groups after gd 15. The number of ovarian corpora lutea and uterine implantation sites, and the percent preimplantation loss were not affected by boric acid exposure.

Offspring body weights were significantly decreased in the 13.3 and 25 mg B/kg-day dose groups on gd 20. The body weight of the low- to high-dose groups, respectively, were 99, 98, 97, 94, and 88% of control weight. There was no evidence of a treatment-related increase in the incidence of external or visceral malformations or variations when considered collectively or individually. On gd 20, skeletal malformations or variations considered collectively showed a significant increased percentage of fetuses with skeletal malformations per litter. Taken individually, dose-related response increases were observed for short rib XIII, considered a malformation in this study, and wavy rib or wavy rib cartilage, considered a variation. Statistical analyses indicated that the incidence of short rib XIII and wavy rib were both increased in the 13.3 and 25 mg B/kg-day dose groups relative to controls. A significant trend test ($p < 0.05$) was found for decrease in rudimentary extra rib on lumbar I, classified as a variation. Only the high-dose group had a biologically relevant, but not statistically significant, decrease in this variation. The LOAEL for Phase I of this study was considered to be 0.1% boric acid (13.3 mg B/kg-day) based on decreased fetal body weight. The NOAEL for

Phase I of this study was considered to be 0.075% boric acid (9.6 mg B/kg-day).

In Phase II, dams were allowed to deliver and rear their litters until postnatal day (pnd) 21. The calculated average doses of boric acid consumed for Phase II dams were 19, 37, 56, 74, and 145 mg/kg-day (3.2, 6.5, 9.7, 12.9, and 25.3 mg B/kg-day). This phase allowed a follow-up period to determine whether the incidence of skeletal defects in control and exposed pups changed during the first 21 postnatal days. Among live born pups, there was a significant trend test for increased number and percent of dead pups between pnd 0 and 4, but not between pnd 4 and 21; this appeared to be due to an increase in early postnatal mortality in the high dose, which did not differ significantly from controls and was within the range of control values for other studies in this laboratory. On pnd 0, the start of Phase II, there were no effects of boric acid on the body weight of offspring (102, 101, 99, 101, and 100% of controls, respectively). There were also no differences through termination on pnd 21; therefore, fetal body weight deficits did not continue into this postnatal period (Phase II). The percentage of pups per litter with short rib XIII was still elevated on pnd 21 in the 0.20% boric acid dose group (25.3 mg B/kg-day), but there was no incidence of wavy rib, and none of the treated or control pups on pnd 21 had an extra rib on lumbar 1. The NOAEL and LOAEL for phase II of this study were 12.9 and 25.3 mg B/kg-day, respectively.

The Institute for Evaluating Health Risks (IEHR, 1997) concluded that there was a consistent correlation between boric acid exposure and the different effects on rib and vertebral development in rats, mice, and rabbits (see the Additional Studies section for effects in mice and rabbits). Of these three species, the rat was the most sensitive to low-dose effects. A causal association between exposure to boric acid and the short rib XIII existed when fetuses were examined at late gestation or when pups where examined at pnd 21. The IEHR (1997) concluded that decreased fetal body weight occurred at the same dose or at doses lower than those at which skeletal changes were observed and that this was the preferred data set for deriving quantitative estimates.

Several benchmark dose (BMD) analyses were conducted (Allen et al., 1996) using all relevant endpoints to analyze data from Heindel et al. (1992) and Price et al. (1996a, 1994) studies alone and combined data from the two studies. Changes in fetal weight were analyzed by taking the average fetal weight for each litter with live fetuses. Those averages were considered to represent variations in a continuous variable. A BMD was defined in terms of a prespecified level of effect, referred to as the benchmark response (BMR) level (Kavlock et al.,

1995). For mean fetal weight analysis, the BMR was a 5% decrease in the mean fetal weight relative to control. The BMDL05 was defined as the 95% lower bound on the dose corresponding to the BMR. A continuous power model was used. Goodness of fit was evaluated using F-tests that compared the lack of model fit to an estimate of pure error.

For all endpoints, the results of the Heindel et al. (1992) and Price et al. (1994, 1996a) studies were compared. The dose-response patterns were examined to determine if a single function could adequately describe the responses in both studies. This determination was based on a likelihood ratio test. The maximum log-likelihoods from the models fit to the two studies considered separately were added together; the maximum log-likelihood for the model fit to the combined results was then subtracted from this sum. Twice that difference is distributed approximately as a chi-square random variable (Cox and Lindley, 1974). The degrees of freedom for that chi-square random variable are equal to the number of parameters in the model plus 1. The additional degree of freedom was available because the two control groups were treated as one group in the combined results, which eliminates the need to estimate one of the intra-litter correlation coefficients (for beta-binomial random variables) or variances (for normal random variables) that was estimated when the studies were treated separately. The critical values from the appropriate chi-square distributions (associated with a p-value of 0.01) were compared to the calculated values. When the calculated value was less than the corresponding critical value, the combined results were used to estimate BMDLs; this result indicated that the responses from the two studies were consistent with a single dose-response function. BMDL05 values calculated with a continuous power model for fetal body weight (litter weight averages) were less than those for all other relevant endpoints. The BMDL05 based on the combined results of the two studies was 10.3 mg B/kg-day, which was very close to the NOAEL of 9.6 mg B/kg-day from the Price et al. (1996a, 1994) study.

In addition to the rat studies, the developmental effects of boric acid were also studied in mice and rabbits. Heindel et al. (1994, 1992) and Field et al. (1989) identified a NOAEL and LOAEL of 43.3 and 79 mg B/kg-day, respectively, for decreased fetal body weight in mice exposed to boric acid in the feed. Increased resorptions and malformations, especially short rib XIII, were noted at higher doses. Price et al. (1996b, 1991) and Heindel et al. (1994) identified a NOAEL and LOAEL of 21.9 and 43.7 mg B/kg-day for developmental effects in rabbits. Frank effects were found at the LOAEL, including high prenatal mortality and increased incidence of malformations, especially cardiovascular defects

VI.A. Oral RfD References

Allen, BC; Strong, PL; Price, CJ; Hubbard, SA; Datson, G.P. (1996) Benchmark dose analysis of developmental toxicity in rats exposed to boric acid. Fund Appl Toxicol 32:194-204.

Anderson, DL; Cunningham, WC; Lindstrom, TR. (1994) Concentrations and intakes of H, B, S, K, Na, Cl, and NaCl in foods. J Food Comp Anal 7:59-82.

Clarke, WB; Gibson, RS. (1988) Lithium, boron and nitrogen in 1-day diet composites and a mixed-diet standard. J Food Comp Anal 1:209-220.

Cox, D; Lindley, D. (1974) Theoretical Statistics. Chapman & Hall, London.

Culver, BD; Hubbard, SA. (1996) Inorganic boron health effects in humans: an aid to risk assessment and clinical judgement. J Trace Elem Exp Med 9:175-184.

Dixon, RL; Sherins, RJ; Lee, IP. (1979) Assessment of environmental factors affecting male fertility. Environ Health Perspect 30:53-68.

Dourson, M; Maier, A; Meek, B; Renwick, A; Ohanian, E; Poirier, K. (1998) Boron tolerable intake re-evaluation of toxicokinetics for data derived uncertainty factors. Biol Trace Elem Research 66(1-3):453-463.

Dunlop, W. (1981) Serial changes in renal haemodynamics during normal human pregnancy. Br J Obstet Gynecol 88:1-9.

ECETOC (European Centre for Ecotoxicology and Toxicology of Chemicals). (1994) Reproductive and General Toxicology of Some Inorganic Borates and Risk Assessment for Human Beings. Technical Report No. 65. Brussels, December.

Fail, PA; George, JD; Seely, JC; Grizzle, TB; Heindel, JJ. (1991) Reproductive toxicity of boric acid in Swiss (CD-1) mice: Assessment using the continuous breeding protocol. Fund Appl Toxicol 17:225-239.

Field, EA; Price, CJ; Marr, MC; Myers, CB; Morrissey, RE. (1989) Final report on the Developmental Toxicity of Boric Acid (CAS No. 10043-35-3) in CD-1-Swiss Mice. NTP Final Report No. 89-250.

National Toxicology Program, U.S. DHHS, PHS, NIH, Research Triangle Park, NC, August 11.

Heindel, JJ; Price, CJ; Field, EA; et al. (1992) Developmental toxicity of boric acid in mice and rats. Fund Appl Toxicol 18:266-277.

Heindel, JJ; Price, CJ; Schwetz, BA. (1994) The developmental toxicity of boric acid in mice, rats and rabbits. Environ Health Perspect 102(Suppl 7):107-112.

Hunt, CD. (1994) The biochemical effects of physiologic amounts of dietary boron in animal nutrition models. Environ Health Perspect 102(Suppl 7):35-43.

IEHR (Institute for Evaluating Health Risks). (1997) An assessment of boric acid and borax using the IEHR evaluative process for assessing human developmental and reproductive toxicity of agents. Reprod Toxicol 11:123-160.

IOM (Institute of Medicine). (2002) Dietary Reference Intakes for Vitamin A, Vitamin K, Arsenic, Boron, Chromium, Copper, Iodine, Iron, Manganese, Molybdenum, Nickel, Silicon, Vanadium and Zinc. National Academy Press, Washington, DC.

Iyengar, GV; Clarke, WB; Downing, RG; Tanner, JT. (1988) Lithium in biological and dietary materials. Proc Intl Workshop, Trace Elem Anal Chem Med Biol 5:267-269.

Kavlock, RJ; Allen, BC; Faustman, EM; Kimmel, CA. (1995) Dose response assessments for developmental toxicity: IV. Benchmark doses for fetal weight changes. Fund Appl Toxicol 26:211-222.

Krutzen, F; Olofsson, P; Back, SE; Nilsson-Ehle, P. (1992) Glomerular filtration rate in pregnancy; a study in normal subjects and in patients with hypertension, preeclampsia and diabetes. Scand J Clin Lab Invest 52:387-392.

Ku, WW; Chapin, RE; Wine, RN; Gladen, BC. (1993) Testicular toxicity of boric acid (BA): Relationship of dose to lesion development and recovery in the F344 rat. Reprod Toxicol 7:305-319.

Linder, RE; Strader, LF; Rehnberg, GL. (1990) Effect of acute exposure to boric acid on the male reproductive system of the rat. J Toxicol Environ Health 31:133-146.

Litovitz, TL; Klein-Schwartz, W; Oderda, GM; Schmitz, BF. (1988) Clinical manifestations of toxicity in a series of 784 boric acid ingestions. Am J Emerg Med 6:209-213.

Mertz, W. (1993) Essential trace metals: new definitions based on new paradigms. Nutr Rev 51:287-295.

Nielsen, FH. (1991) Nutritional requirements for boron, silicon, vanadium, nickel, and arsenic: Current knowledge and speculation. FASEB J 5:2661-2667.

Nielsen, FH. (1992) Facts and fallacies about boron. Nutr Today 27:6-12.

Nielsen, FH. (1994) Biochemical and physiologic consequences of boron deprivation in humans. Environ Health Perspect 102(Suppl. 7):59-63.

Nielsen, FH; Hunt, CD; Mullen, LM; Hunt, JR. (1987) Effect of dietary boron on minerals, estrogen, and testosterone metabolism in post-menopausal women. FASEB J 1:394-397.

NRC (National Research Council). (1989) Recommended Dietary Allowances, 10th ed. National Academy Press, Washington, DC. p. 267.

NTP (National Toxicology Program). (1987) Toxicology and Carcinogenesis Studies of Boric Acid (CAS No. 10043-35-3) in B6C3F1 Mice (feed studies). NTP Tech. Rep. Ser. No. 324. U.S. DHHS, PHS, NIH, Research Triangle Park, NC.

Pahl, MV; Culver, BD; Strong, PL; Murray, FJ; Vaziri, ND. (2001) The effect of pregnancy on renal clearance of boron in humans: a study based on normal dietary intake of boron. Toxicol Sci 60(2):252-256.

Price, CJ; Field, EA; Marr, MC; Myers, CB; Morrissey, RE; Schwetz, BA. (1990) Final report on the Developmental Toxicity of Boric Acid (CAS No. 10043-35-3) in Sprague Dawley Rats. NTP Report No. 90-105 (and Report Supplement No. 90-105A). National Toxicology Program, U.S. DHHS, PHS, NIH, Research Triangle Park, NC, May 1.

Price, CJ; Marr, MC; Myers, CB; Heindel, JJ; Schwetz, BA. (1991) Final report on the Developmental Toxicity of Boric Acid (CAS No. 10043-35-3) in New Zealand White Rabbits. NTP TER-90003. National Toxicology Program, U.S. DHHS, PHS, NIH, Research

Triangle Park, NC, November (and Laboratory Supplement No. TER-90003, December).

Price, CJ; Marr, MC; Myers, CB. (1994) Determination of the No-Observable-Adverse-Effect Level (NOAEL) for Developmental Toxicity in Sprague-Dawley (CD) Rats Exposed to Boric Acid in Feed on Gestational Days 0 to 20, and Evaluation of Postnatal Recovery through Postnatal Day 21. Final report. (3 volumes, 716 pp). RTI Identification No. 65C-5657-200. Research Triangle Institute, Center for Life Science, Research Triangle Park, NC.

Price, CJ; Strong, PL; Marr, MC; Myers, CB; Murray, FJ. (1996a.) Developmental toxicity NOAEL and postnatal recovery in rats fed boric acid during gestation. Fund Appl Toxicol 32:179.

Price, CJ; Marr, MC; Myers, CB; Seely, JC; Heindel, JJ; Schwetz, BA. (1996b) The developmental toxicity of boric acid in rabbits. Fund Appl Toxicol 34:176-187.

Schou, JS; Jansen, JA; Aggerbeck, B. (1984) Human pharmacokinetics and safety of boric acid. Arch Toxicol 7:232-235.

Seal, BS; Weeth, HJ. (1980) Effect of boron in drinking water on the male laboratory rat. Bull Environ Contam Toxicol 25:782-789.

Sturgiss, SN; Wilkinson, R; Davison, JM. (1996) Renal reserve during human pregnancy. Am J Physiol 271:F16-F20.

Treinen, KA; Chapin, RE. (1991) Development of testicular lesions in F344 rats after treatment with boric acid. Toxicol Appl Pharmacol 107:325-335.

U.S. Borax Research Corp. (1963) MRID No. 00068026; HED Doc. No. 009301. Available from EPA. Write to FOI, EPA, Washington, DC. 20460.

U.S. Borax Research Corp. (1966) MRID No. 00005622, 00068021, 00068881; HED Doc. No. 009301. Available from EPA. Write to FOI, EPA, Washington, DC. 20460.

U.S. Borax Research Corp. (1967) MRID No. 00005623, 005624; HED Doc. No. 009301. Available from EPA. Write to FOI, EPA, Washington, DC. 20460.

U.S. Borax. (2000) UCI Boric Acid Clearance Study Reports and Associated Data: Rat and Human Studies, April 4, 2000.

U.S. EPA. (1998) Science Policy Council Handbook: Peer Review. Prepared by the Office of Science Policy, Office of Research and Development, Washington, DC. EPA 100-B-98-001.

U.S. EPA. (1999) Guidelines for Carcinogen Risk Assessment. Revised Draft. Risk Assessment Forum, Washington, DC. July 1999. Available online from: http://www.epa.gov/cancerguidelines/draft-guidelines-carcinogen-ra-1999.htm

U.S. EPA. (2004) Toxicological Review of Boron and Compounds in Support of Summary Information on Integrated Risk Information (IRIS). National Center for Environmental Assessment, Washington, DC. Available online from: http://www.epa.gov/iris.

Vanderpool, RA; Hof, D; Johnson, PE. (1994) Use of inductively coupled plasma-mass spectrometry in boron-10 stable isotope experiments with plants, rats, and humans. Environ Health Perspect 102(Suppl 7):13-20.

Vaziri, ND; Oveisi, F; Culver, BD; Pahl, MV; Andersen, ME; Strong, PL; Murray, FJ. (2001) The effect of pregnancy on renal clearance of boron in rats given boric acid orally. Toxicol Sci 60(2):257-263.

Weir, RJ; Fisher, RS. (1972) Toxicologic studies on borax and boric acid. Toxicol Appl Pharmacol 23:351-364.

Appendix 14

Manganese in Drinking Water: Study Suggests Adverse Effects On Children's Intellectual Abilities
Science News: ScienceDaily (Sep. 20, 2010)
http://www.sciencedaily.com/releases/2010/09/100920074013.htm

Their results are published in the journal Environmental Health Perspectives, in an article entitled "Intellectual Impairment in School-Age Children Exposed to Manganese from Drinking Water."

Manganese: toxic in the workplace but harmless in water?

The neurotoxic effects of manganese exposure in the workplace are well known. This metal is naturally occurring in soil and in certain conditions is present in groundwater. In several regions of Quebec and Canada and in other parts of the world, the groundwater contains naturally high levels of manganese. Does it pose a danger? What effect might it have on children's health? This is the first study to focus on the potential risks of exposure to manganese in drinking water in North America.

The study, carried out by researchers at the Université du Québec à Montréal, the Université de Montréal and the École Polytechnique de Montréal, examined 362 Quebec children, between the ages of 6 and 13, living in homes supplied by with groundwater (individual or public wells). For each child, the researchers measured the concentration of manganese in tap water from their home, as well as iron, copper, lead, zinc, arsenic, magnesium and calcium. The amount of manganese from both tap water and food was estimated from a questionnaire. Finally, each child was assessed with a battery of tests assessing cognition, motor skills, and behaviour.

Lead author Maryse Bouchard explains, "We found significant deficits in the intelligence quotient (IQ) of children exposed to higher concentration of manganese in drinking water. Yet, manganese concentrations were well below current guidelines." The average IQ of children whose tap water was in the upper 20% of manganese concentration was 6 points below children whose water contained little or no manganese. The analyses of the association between manganese in tap water and children's IQ took into account various factors such as family income, maternal intelligence, maternal education, and the presence of other metals in the water. For co-author Donna Mergler, "This is a very marked effect; few environmental contaminants have shown such a strong correlation with intellectual ability." The authors state that the amount of manganese present in food showed no relationship to the children's IQ.

What next?

So what can be done about it? Some of the municipalities where the study was conducted have already installed a filtration system that removes manganese from the water. According to one of the other co-authors of the study, Benoit Barbeau, NSERC Industrial Chair in Drinking Water at the École Polytechnique de Montréal, "A viable alternative solution is home use of filtering pitchers that contain a mixture of resins and activated carbon. Such devices can reduce the concentration of manganese by 60% to100% depending on filter use and the characteristics of the water."

In Quebec, where the study was conducted, manganese is not on the list of inorganic substances in the Ministry of Sustainable Development, Environment and Parks Regulation respecting the quality of drinking water. "Because of the common occurrence of this metal in drinking water and the observed effects at low concentrations, we believe that national and international guidelines for safe manganese in water should be revisited." the authors conclude.

Chapter 13 – Water quality after sewage spill

The question asked, what would raw sewage do to water quality in Winnipeg, Manitoba, Canada [see **Appendix 15a & 15b**, cited below].

It would also be important to know what the water quality was after this raw sewage spill:

"More than one billion litres of partially treated sewage have flowed into the Red River over the past few weeks, due to a major glitch at a Winnipeg waste treatment plant, according to city officials."[1]

"One of the four sewage treatment processes at the South End Water Pollution Control Plant has not been working since Oct. 7, and city staff are stumped as to how and why it broke down."[2]

"One of Winnipeg's sewage treatment plants has been spewing wastewater that contains up to 10 times the recommended amount of fecal coliform bacteria into the Red River for nearly a month."[3]

"The city is licensed to keep the count below 200 bacteria per 100 ml of water, and current levels are between 1,000 and 2,000, city officials said."[4]

"Over two-million litres of untreated sewage was discharged into the Assiniboine River last month over a five-day period."[5]

"It was the largest sewage overflow in Winnipeg since the city's massive sewage spill in 2002."[6]

"That's on top of 17 smaller spills that occurred this year between March and April."[7]

"It's all due to Winnipeg's outdated combined sewer system, which diverts raw sewage into our rivers every time it rains, during spring runoff and when pipes get clogged, like they did last month."[8]

The current authors ironically had had their tap water in Winnipeg, Manitoba, Canada tested after the raw sewage spill on 24 November, 2011.

The results of this tap water test appear in **Figure 15**, appearing below.

Figure 15: Water quality after the raw sewage spill

Sample Details/Parameters	Result	Qualifier*	D.L.	Units	Extrac
Total Coliform and E.coli					
Total Coliforms	0		0	MPN/100mL	
Escherichia Coli	0		0	MPN/100mL	
Miscellaneous Parameters					
Fecal Coliforms	<3		3	MPN/100mL	
Total Metals by ICP-MS					
Boron (B)-Total	<0.030		0.030	mg/L	24-NOV
Calcium (Ca)-Total	17.9		0.20	mg/L	24-NOV
Copper (Cu)-Total	0.0421		0.0020	mg/L	24-NOV
Iron (Fe)-Total	0.13		0.10	mg/L	24-NOV
Magnesium (Mg)-Total	5.13		0.050	mg/L	24-NOV
Manganese (Mn)-Total	0.0150		0.0010	mg/L	24-NOV
Potassium (K)-Total	1.08		0.10	mg/L	24-NOV
Sodium (Na)-Total	31.1		0.050	mg/L	24-NOV
Zinc (Zn)-Total	<0.020		0.020	mg/L	24-NOV
WA5					
Alkalinity					
Alkalinity, Total (as CaCO3)	66.1		1.0	mg/L	
Bicarbonate (HCO3)	80.6		2.0	mg/L	
Carbonate (CO3)	<0.60		0.60	mg/L	
Hydroxide (OH)	<0.40		0.40	mg/L	
Chloride					
Chloride	21.6		0.50	mg/L	
Conductivity					
Conductivity	281		20	umhos/cm	
Nitrate as N					
Nitrate-N	<0.050		0.050	mg/L	
Nitrate+Nitrite					
Nitrate and Nitrite as N	<0.071		0.071	mg/L	
Nitrite as N					
Nitrite-N	<0.050		0.050	mg/L	
Phosphorus, Total					
Phosphorus (P)-Total	0.582		0.010	mg/L	
Sodium Adsorption Ratio					
Sodium Adsorption Ratio	1.67		0.030		
Sulfate					
Sulfate	49.1		0.50	mg/L	
pH					
pH	7.68		0.10	pH units	

In examining the results cited before the raw sewage spill [**Figure 9**] and after the raw sewage spill [**Figure 15**], there doesn't appear to be an appreciable difference in results.

A comparison of results appears in **Table 1**, appearing below.

Table 1

Tap Water Test Date	10 August, 2011	24 November, 2011
Sodium absorption rate	-	1.67
Calcium (Ca)	24.8 mg/L	17.9 mg/L
Magnesium (Mg)	6.63 mg/L	5.13 mg/L
Sodium	34.3 mg/L	31.1 mg/L
Ph	7.98 pH units	7.68 pH units
Total dissolved oxygen	190 mg/L	-
Conductivity	293 umhos/cm	281 umhos/cm
Chloride	23.4 mg/L	21.6 mg/L
Sulfate	55.1 mg/L	49.1 mg/L
Phosphorous	0.612 mg/L	0.582 mg/L
Potassium	1.48 mg/L	1.08 mg/L
Bicarbonate	-	80.6 mg/L
Alkalinity	-	66.1 mg/L
Escherichia coli	< 1	0
Heterotrophic plate count Fecal coliforms	< 10	< 3
Total coliforms	< 1	0

With respect to **Calcium**, it was reported by Morr *et. al.* (2006) that there was "substantial variability in tap water calcium concentrations ... in the USA, tap water calcium concentrations varied from 8.3 mg/L in Montgomery, AL, to 131 mg/L in Phoenix, AZ."[1]

"The average calcium concentration was 50.6 ± 29.4 mg/L, and the percentage of RDI of calcium satisfied by six 8-oz glasses per day varied from 0.85 to 13.5%, with an average of 5.2%."[2]

"There was no statistically significant difference between regions."[3]

In addition, it was mentioned by Morr *et. al.* (2006), "Canadian cities were equally diverse: the tap water calcium concentrations varied from 1.4 mg/L in Vancouver to 135.5 mg/L in Kitchener."[4]

" The corresponding RDI percentages varied from 0.15 to 13.9%. The average for Canada was 48.8 ± 53.2 mg/L."[5]

With respect to **Magnesium**, as cited by Durlach *et. al.* (1989), "of all the cardiovascular risk factors, magnesium now takes first place as judged by the accumulation of epidemiological, pathophysiological, clinical and experimental data, both pharmacological and therapeutic."[6]

In addition, as reported, "Magnesium is the fourth most abundant cation in the body and the second most abundant cation in intracellular fluid. It is a cofactor for some 350 cellular enzymes, many of which are involved in energy metabolism."[7]

"It is also involved in protein and nucleic acid synthesis and is needed for normal vascular tone and insulin sensitivity. Low magnesium levels are associated with endothelial dysfunction, increased vascular reactions, elevated circulating levels of C- reactive protein and decreased insulin sensitivity. Low magnesium status has been implicated in hypertension, coronary heart disease, type 2 diabetes mellitus and the metabolic syndrome."[8]

In addition, "based on identified case–control and cohort studies, there is no evidence of an association between water hardness or calcium and acute myocardial infarction or deaths from cardiovascular disease (acute myocardial infarction, stroke and hypertension). There does not appear to be an association between drinking-water magnesium and acute myocardial infarction."[9]

"However, the studies do show a negative association (i.e. protective effect) between cardiovascular mortality and drinking-water magnesium. Although this association does not necessarily demonstrate causality, it is consistent with the well known effects of magnesium on cardiovascular function."[10]

With respect to **Sodium**, as reported "the taste threshold for sodium in water depends on the associated anion and the temperature of the solution. At room temperature, the threshold values are about 20 mg/litre for sodium carbonate, 150 mg/litre for sodium chloride, 190 mg/litre for sodium nitrate, 220 mg/litre for sodium sulfate, and 420 mg/litre for sodium bicarbonate."[11]

"Sodium salts are generally highly soluble in water and are leached from the terrestrial environment to groundwater and surface water. They are non-volatile and will thus be found in the atmosphere only in association with particulate matter."[12]

"The sodium ion is ubiquitous in water. Most water supplies contain less than 20 mg of sodium per litre, but in some countries levels can exceed 250 mg/litre. Saline intrusion, mineral deposits, seawater spray, sewage effluents, and salt used in road de-icing can all contribute significant quantities of sodium to water. In addition, water-treatment chemicals, such as sodium fluoride, sodium bicarbonate, and sodium hypochlorite, can together result in sodium levels as high as 30 mg/litre. Domestic water softeners can give levels of over 300 mg/litre, but much lower ones are usually found."[13]

"In a survey of 2100 water samples in the USA in 1963–1966, the sodium ion concentrations found were in the range 0.4–1900 mg/litre; in 42% of the samples, the concentrations were in excess of 20 mg/litre, but in 5% they were greater than 250 mg/litre."[14]

"In a later survey of 630 water-supply systems in the same country, the sodium ion concentrations found ranged from less than 1 to 402 mg/litre, with similar distribution of values."[15]

"Sodium salts are found in virtually all food (the main source of daily exposure) and drinking - water. Sodium levels in the latter are typically less than 20 mg/litre but can markedly exceed this in some countries. On the basis of existing data, no firm conclusions can be drawn concerning the possible association between sodium in drinking-water and the occurrence of hypertension. No health-based guideline value is therefore proposed. However, sodium may affect the taste of drinking-water at levels above about 200 mg/litre."[16]

With respect to **Conductivity**, as noted "is a measure of the ability of water to pass an electrical current. Conductivity in water is affected by the presence of inorganic dissolved solids such as chloride, nitrate, sulfate, and phosphate anions (ions that carry a negative charge) or sodium, magnesium, calcium, iron, and aluminum cations (ions that carry a positive charge)."[17]

"Organic compounds like oil, phenol, alcohol, and sugar do not conduct electrical current very well and therefore have a low conductivity when in water. Conductivity is also affected by temperature: the warmer the water, the higher the conductivity. For this reason, conductivity is reported as conductivity at 25 degrees Celsius (25 C)."[18]

"The basic unit of measurement of conductivity is the mho or siemens. Conductivity is measured in micromhos per centimeter (μmhos/cm) or microsiemens per centimeter (μs/cm)."[19]

"Distilled water has a conductivity in the range of 0.5 to 3 μmhos/cm."[20]

"The conductivity of rivers in the United States generally ranges from 50 to 1500 μmhos/cm."[21]

"Studies of inland fresh waters indicate that streams supporting good mixed fisheries have a range between 150 and 500 μhos/cm."[22]

"Conductivity outside this range could indicate that the water is not suitable for certain species of fish or macroinvertebrates. Industrial waters can range as high as 10,000 μmhos/cm."[23]

With respect to **Chloride**, as cited "a normal adult human body contains approximately 81.7 g chloride. On the basis of a total obligatory loss of chloride of approximately 530 mg/day, a dietary intake for adults of 9 mg of chloride per kg of body weight has been recommended (equivalent to slightly more than 1 g of table salt per person per day). For children up to 18 years of age, a daily dietary intake of 45 mg of chloride should be sufficient."[24]

"A dose of 1 g of sodium chloride per kg of body weight was reported to have been lethal in a 9-week-old child."[25]

"Chloride toxicity has not been observed in humans except in the special case of impaired sodium chloride metabolism, e.g. in congestive heart failure."[26]

"Healthy individuals can tolerate the intake of large quantities of chloride"; however, "little is known about the effect of prolonged intake of large amounts of chloride in the diet."[27]

"In experimental animals, hypertension associated with sodium chloride intake appears to be related to the sodium rather than the chloride ion."[28]

"Chloride concentrations in excess of about 250 mg/litre can give rise to detectable taste in water, but the threshold depends upon the associated cations. Consumers can, however, become accustomed to concentrations in excess of 250 mg/litre. No health-based guideline value is proposed for chloride in drinking-water."[29]

With respect to **Sulfate**, as reported, "Sulfate (SO_4^{2-}) is a divalent anion. There may be up to one percent sulfate present in gastric fluids. The human body distinguishes sulfate (5.7 angstroms) from phosphate (6 angstroms) and thiosulfate ($S_2O_3^{2-}$) by the distinctive stereospecific molecular configurations of each ion."[30]

"The body maintains a homeostasis between absorbed inorganic sulfate, sulfate compounds, and renal excretion; membrane transport and regulation contribute to this homeostasis."[31]

"Inorganic sulfate represents a small fraction of total sulfate in the body, which includes muccopolysaccharides, chondroitin sulfate, glycolipids, steroids, thyroid hormones, peptide hormones (gastrin), oligosaccharides, and xenobiotics (e.g., drugs). As much as 5 to 10 percent of excreted sulfate is excreted as sulfyl esters."[32]

"There are very few scientific reports describing the health effects of exposure to sulfate in drinking water, and the concerns regarding sensitive populations are based on case studies and anecdotal reports."[33]

"One such potentially sensitive population is infants receiving their first bottles containing tap water, either as water alone or as formula mixed with water."[34]

"Another group of people who could potentially be adversely affected by water with high sulfate concentrations are transient populations (i.e., tourists, hunters, students, and other temporary visitors) and people moving into areas with high sulfate concentrations in the drinking water from areas with low sulfate concentrations in drinking water."[35]

With respect to **Phosphorous**, as cited "Nitrogen and phosphorus levels vary naturally and some amount in a water body is not harmful."[36]

"In fact, nitrogen and phosphorus are essential to maintain the health of the organisms that live there. When too much nitrogen and phosphorus enter surface waters; however, they cause the ecosystem to become unbalanced."[37]

"Nitrogen and phosphorus are often a result of human activities and they speed up the growth of algae in surface waters to an unhealthy level in a process called eutrophication. The algae grow out of control and form what is called an algal bloom. These algal blooms can cause many problems for underwater plants and animals, as well as humans."[38]

"Nitrogen and phosphorus (along with co-contaminants such as pathogens, chemicals, and animal pharmaceuticals) are also found in excess in ground water, which some homes use as their drinking water source."[39]

"At levels above the 10 mg/L maximum contaminant level (MCL) in ground water, nitrates can cause human health effects, such as blue baby syndrome. To protect human health, it is important that nitrate levels be below this 10 mg/L MCL when the ground water is used as drinking water."[40]

With respect to **Potassium**, as noted "is a dietary requirement for us, and we take up about 1-6 g per day at a requirement of 2-3.5 g per day."[41]

"The total potassium amount in the human body lies somewhere between 110 and 140 g and mainly depends upon muscle mass. The muscles contain most potassium after red blood cells and brain tissue."[42]

"Whereas its opponent sodium is present in intracellular fluids, potassium is mainly present within cells. It preserves osmotic pressure. The relation of potassium in cells to potassium in plasma is 27:1, and is regulated by means of sodium-potassium pumps."[43]

"Vital functions of potassium include its role in nerve stimulus, muscle contractions, blood pressure regulation and protein dissolution."[44]

"It protects the heart and arteries, and may even prevent cardiovascular disease."[45]

"Potassium shortages are relatively rare, but may lead to depression, muscle weakness, heart rhythm disorder and confusion. Potassium loss may be a consequence of chronic diarrhoea or kidney disease, because the physical potassium balance is regulated by the kidneys."[46]

In addition, as noted, "adverse health effects due to potassium consumption from drinking-water are unlikely to occur in healthy individuals. Potassium intoxication by ingestion is rare, because potassium is rapidly excreted in the absence of pre-existing kidney damage and because large single doses usually induce vomiting."[47]

"Although potassium may cause some health effects in susceptible individuals, potassium intake from drinking-water is well below the level at which adverse health effects may occur."[48]

"Health concerns would be related to the consumption of drinking - water treated by potassium-based water treatment (principally potassium chloride for regeneration of ion exchange water softeners), affecting only individuals in the high - risk groups (individuals with kidney dysfunction or other diseases such as heart disease, coronary artery disease, hypertension, diabetes, adrenal insufficiency, pre-existing hyperkalaemia; people taking medications that interfere with normal potassium - dependent functions in the body; and older individuals or infants). It is recommended that susceptible individuals seek medical advice to determine whether they should avoid the consumption of water (for drinking or cooking) treated by water softeners using potassium chloride."[49]

With respect to **Escherichia coli**, **Heterotrophic plate count**, **Fecal coliforms** and **Total coliforms** there apparently was no adverse effect from the 5 billion litre raw sewage spill in Winnipeg, Manitoba, Canada [see **Table 1**, cited above].

With respect to raw sewage spills [see **Appendix 16a & 16b** cited below], however, it is mentioned that "raw sewage contains biological agents such as bacteria, viruses, fungi and parasites that can cause serious illness and even death. There is also a risk from contamination with unknown chemicals (such as solvents, carcinogens, pesticides) and from toxic, irritant, asphyxiating or flammable gases in confined spaces."[50]

And, "always assume that floodwater is contaminated with sewage. Immediate clean-up is essential to reduce the risk of infection and/or mould growth."[51]

In addition, "Tetanus is caused by a toxin produced by the bacterium Clostridium tetani that is common in soil and in sewage. The bacterium enters the body via open wounds. There is a high risk of death occurring if infected. Anyone who may be exposed to sewage or soil should have prophylaxis tetanus vaccinations every ten years."[52]

"Leptospirosis is caused by the parasitic worm Leptospira icterohaemorrhegiae and is transmitted from water and damp earth contaminated primarily by rats that harbour the organism. The initial septicemia phase lasts for 4-7 days and causes acute headache, chills, fever, severe muscle aching, anorexia, nausea and vomiting. The immune phase, characterised by aseptic meningitis, follows a 24-72 hour asymptomatic period. Approximately 10-15% of patients present with Weil's disease, jaundice, hemorrhage and renal damage."[53]

"Hepatitis A is caused by the Hepatitis A virus (HAV) that is transmitted primarily by ingestion. The virus must be present in sufficient quantities to cause infection. Infection occurs after an incubation period of three to four weeks. Hepatitis A is often mild, but can be severe or even fatal in some cases. Symptoms are fever, headache, nausea and pain in the abdomen, dark urine and jaundice. People can spread the disease to others in the immediate period before they become ill and while they are ill. Recovery from Hepatitis A can be slow and require several weeks or months of increased rest. A majority of patients make a complete recovery but the disease can be more severe in older patients."[54]

"Giardia and Cryptosporidium are protozoan parasites, commonly found in sewage and surface waters, that can cause diarrhea, stomach cramps, nausea and sometimes fever. Symptoms may last for only a few days or can last for months or years. Many people, especially children, have no symptoms. Cysts from infected persons or animals enter sewage and if untreated may infect other people who ingest the cysts."[55]

"Gram-negative bacteria such as E.coli can cause gastro-intestinal diseases if ingested or airway problems, headache, tiredness and nausea if inhaled. Substances called endotoxins that are released at the time of death of the bacterium have been suggested as the cause of a wide variety of occupational diseases such as mill fever and grain fever."[56]

As also reported, "microbes in raw sewage can enter the body via the nose, mouth, open wounds or by inhalation of aerosols or dusts. The most common modes of infection are through drinking contaminated water or hand to mouth transmission. Skin contact alone does not pose a health threat unless you have an open wound."[57]

"The survival of pathogens depends on a number of factors: location, type of surface contaminated, whether disinfectants are used and environmental conditions. UV radiation reduces the survival rate of pathogens. Mild temperatures and higher humidity increase survival times. The risk of exposure when handling sewage can be reduced significantly by effective and immediate clean-up and by taking appropriate safety precautions."[58]

Footnotes

1 – 5. Simon Morr, Esteban Cuartas, Basil Alwattar, and Joseph M. Lane, *How Much Calcium Is in Your Drinking Water? A Survey of Calcium Concentrations in Bottled and Tap Water and Their Significance for Medical Treatment and Drug Administration*, HSS J. 2006 September; 2(2): 130–135.
PMCID: PMC2488164
http://www.ncbi.nlm.nih.gov/pmc/articles/PMC2488164/

6. J. Durlach, M. Bara and A. Guiet-Bara, "Magnesium level in drinking water: its importance in cardiovascular risk". In: *Magnesium in Health and Disease.* Y. Itokawa and J. Durlach eds. John Libbey. London, 1989; 173-182

Also see: Durlach, J. (1988): *Magnesium in clinical practice*, Publ. John Libbey, pp. 386 London-Paris.

Also see: Durlach, J., Bara, M., Guiet-Bara, A. (1985): *Magnesium level in drinking water and cardiovascular risk factor: a hypothesis.* Magnesium 4, 5-15.

7 - 10. *WHO Meeting of Experts on the Possible Protective Effect of Hard Water Against Cardiovascular Disease Washington, D.C., USA 27–28 April 2006*
Public Health and Environment World Health Organization
WHO/SDE/WSH/06.06
Geneva 2006
https://docs.google.com/viewer?a=v&q=cache:lXKG6tC-L5AJ:www.who.int/water_sanitation_health/gdwqrevision/cardiofullreport.pdf+inverse+relationship+between+various+types+of+heart+diseases+and+magnesium&hl=en&gl=uk&pid=bl&srcid=ADGEESjnyotRvGdK2OtsHgbpS5Ko8H__5zq5AGSwUF85LKesTrNwDl9m-NHGCDYGKi3TcoXGs0-FHMR7tkcT5_PGAocXxKGuDiuc0TtCV65k-w3nQYU1E0PqAF_g_8TQXwvBWzYqG9cO&sig=AHIEtbQeudzgpUZr7d24F09_LxcyJqwq0w

11 - 16. *Sodium in Drinking-water*
Background document for development of WHO Guidelines for Drinking-water Quality
WHO/SDE/WSH/03.04/15
https://docs.google.com/viewer?a=v&q=cache:t7UxjGO5qW8J:www.who.int/water_sanitation_health/dwq/chemicals/sodium.pdf+sodium+in+drinking+water+health+effects&hl=en&gl=ca&pid=bl&srcid=ADGEESjgynX4ImERwW3FEzlpIrwCR5m_Jvvi_lwSvVRsIQnxm6V8NVIHfOgw_YgKJnIhDiam_2TJWqBaVCI0UH-7r3T2Z5OfckzrJVZEbW34DftfHxxAQcVUG4eU_PpcZF5Jj1QIO3DC&sig=AHIEtbT6v14_eH6mxE80Dw6oirE9GYTmfQ

Also see: *Sodium, chlorides and conductivity in drinking water.* Copenhagen, WHO Regional Office for Europe, 1979 (EURO Reports and Studies No. 2).

Also see: National Academy of Sciences. *Drinking water and health.* Washington, DC, National Academy Press, 1977:400-411.

17 - 23. *EPA 5.9 Conductivity: What is conductivity and why is it important?*
http://water.epa.gov/type/rsl/monitoring/vms59.cfm

24 - 29. *Chloride in Drinking-water*

Background document for development WHO Guidelines for Drinking-water
Quality
WHO/SDE/WSH/03.04/03
https://docs.google.com/viewer?a=v&q=cache:3Izf4enUwBIJ:www.who.int/water_
sanitation_health/dwq/chloride.pdf+chloride+in+drinking+water+health+effects&hl
=en&gl=ca&pid=bl&srcid=ADGEEShJRQkrbQPzqJzZvEKYnEOvRBusaeE2OrgP
U6JQG4AqBHMJIiFSeg1PMS0O5IDh_V_LBUuGIl2fep_h6AuAxYc50RGcHdq0
KeEH3psw7HsQlYvTuPgpb2L1HmVDKtWGN5vXkydL&sig=AHIEtbS4jFjvIVY
RPXPRoC64iIR38jHx7w

Also see: Department of National Health and Welfare (Canada). *Guidelines for Canadian drinking water quality*. Supporting documentation. Ottawa, 1978.

Also see: *Sodium, chlorides, and conductivity in drinking water: a report on a WHO working group*. Copenhagen, WHO Regional Office for Europe, 1978 (EURO Reports and Studies 2).

Also see: Wesson LG. *Physiology of the human kidney*. New York, NY, Grune and Stratton, 1969: 591

30 - 35. *Health Effects from Exposure to Sulfate in Drinking Water Workshop*
United States Environmental Protection Agency: Office of Water 4607
EPA 815-R-99-002, January 1999
https://docs.google.com/viewer?a=v&q=cache:s_yVniLGskoJ:www.epa.gov/ogwd
w/contaminants/unregulated/pdfs/summary_sulfate_epa-
cdcworkgroup.pdf+sulfate+in+drinking+water+health+effects&hl=en&gl=ca&pid=
bl&srcid=ADGEESi65mWek5SihQWPfM0HIbKHQdmanujbzHiASn3YQyyU1spS
caAbHuEEVCKOUfGiU7SRhFDwv9ubSBnKYXsv6ysLMDv2bbAcXcvsdHKgeF
Hb-1GzM8oIK7Dz7J8FNeavOH-
pb0U9&sig=AHIEtbRw6VFWpNJ3OE2fmwooG1Li5irFqg

36 – 40. *EPA: Nitrogen and Phosphorus and Healthy Ecosystems*
http://water.epa.gov/scitech/swguidance/standards/criteria/nutrients/problem.cfm

41 - 46. *Lenntech: Potassium (K) and water - Potassium and water: reaction mechanisms, environmental impact and health effects*
http://www.lenntech.com/about/aboutlenntech-en.htm

47 - 49. *Potassium in drinking-water*
Background document for development of WHO Guidelines for Drinking-water
Quality
WHO/HSE/WSH/09.01/7
https://docs.google.com/viewer?a=v&q=cache:J2emdEThHsQJ:whqlibdoc.who.int/
hq/2009/WHO_HSE_WSH_09.01_7_eng.pdf+potassium+in+drinking+water+healt
h+effects&hl=en&gl=ca&pid=bl&srcid=ADGEEShtjPS2RUR0zDxc9DBzVIJ-
YbvSKpi6HSk72bxqJJw1B5lq_IvlE8_wUlJWbJ4tp2BIOzrXk1F6lok2VsH_LIeO4

ktfzfJ_dvRZQhQo8zyKbnh93x5eL83_3oW_lsBtHSMTVtSq&sig=AHIEtbQK_Ft
m0lz__ANT9m3qQfPuKpUAoA

Also see: Gosselin RE, Smith RP, Hodge HC (1984) *Clinical toxicology of commercial products*, 5th ed. Baltimore, MD, Williams & Wilkins.

50 - 58. *Sewage Spills*
Workers Health Centre, 2004
http://www.workershealth.com.au/facts042.html

Appendix 15a

Partially treated sewage flows from Winnipeg plant
CBC News
Posted: Nov 2, 2011 4:36 PM CT
http://www.cbc.ca/news/canada/manitoba/story/2011/11/02/winnipeg-wastewater-plant-problem.html

More than one billion litres of partially treated sewage have flowed into the Red River over the past few weeks, due to a major glitch at a Winnipeg waste treatment plant, according to city officials.

One of the four sewage treatment processes at the South End Water Pollution Control Plant has not been working since Oct. 7, and city staff are stumped as to how and why it broke down.

As a result of the malfunction, effluent coming from the plant — which flows into the Red River — is currently being treated to just 50 per cent of how it would normally be treated, officials said Wednesday.

"Right now we are discharging wastewater to the Red River from the effluent of the plant that is in excess of our license requirement," Mike Shkolny, the city's manager of engineering services, told reporters.

Shkolny said it may sound like a lot of sewage is flowing into the Red River, but he stressed that it's actually a relatively small amount.

He added that it's not raw sewage going into the river, but partially treated sewage.

Biological stage not working

What is malfunctioning is the biological treatment stage, in which microorganisms eat organic material in the sewage.

"For an unknown reason, these microorganisms suddenly stopped thriving on Oct. 7, 2011, upsetting the full biological treatment process," the city stated in a release.

The first two treatment stages — in which grit, sediment and grease are removed from the waste — are working normally.

However, the final ultraviolet disinfection stage is operating at "reduced effort" because the biological treatment stage isn't working, officials say.

Shkolny said city engineers have been unsuccessful in trying to fix the problem so far.

A team of experts has been assembled to work on the issue, but it could be at least another month before they figure out why the treatment process broke down.

Does not meet provincial standards

City officials say this is the first time such a disruption has happened in the history of Winnipeg's sewage treatment plants.

The South End plant, which opened in 1974, treats 60 million litres of sewage a day.

Considering the sewage has been partially treated for the past 26 days, that means more than one billion litres have gone into the river to date.

Shkolny said the problem is not a public health issue, but provincial regulators have been notified because the partially treated effluent does not meet Environment Act licence requirements for the plant.

Fewer people are going in and around the Red River at this time of year, but those who come into contact with the river water should wash their hands thoroughly afterward, he said.

"Full-body contact immersion would not be recommended," Shkolny said.

"However, at this point in time, there's not much recreation going on in the river, so the risk to public health relative to swimming or boating or water-skiing is quite small."

Fishing outfitter surprised

Anyone who catches fish from the river should wash and boil their catch before eating it, he added.

Stu McKay, who owns a fishing outfitting store in nearby Lockport, Man., said the plant malfunction is an example of people's disrespect for water.

McKay said he was surprised to learn that partially treated sewage has been going into the Red River, which runs near his store, Cats on the Red.

"It's amazing, actually, in this day and age that we don't have systems put in place to prevent these types of things [from] happening. I mean, what does it take?" McKay told CBC News.

"Do we not respect water, or should we not be giving it more respect than what we have been in this day and age, knowing that it's the most important resource we have on the planet?"
Make sewers election issue 14

Appendix 15b

Sewage treatment plant spewing into Red
By Paul Turenne, Winnipeg Sun
First posted: Wednesday, November 02, 2011
http://www.winnipegsun.com/2011/11/02/sewage-treatment-plant-
spewing-into-red

*One of Winnipeg's sewage treatment plants has been spewing
wastewater that contains up to 10 times the recommended amount of
fecal coliform bacteria into the Red River for nearly a month.*

*The problem, which one city official called "a major upset," was first
detected Oct. 7, and since then has allowed between 50 million and 60
million litres of only partly treated wastewater to flow into the river
every day. Officials are baffled as to the cause, and say it could be at
least a month before the problem is completely fixed.*

*"This would be a major upset and we're having trouble bringing the
plant back to normal operation," said Kelly Kjartanson, manager of
environmental standards for the city.*

*"If you come into contact with the river, wash your hands," said Mike
Shkolny, manager of the city's engineering division.*

*The plant in question is the south end sewage treatment plant, located
near St. Mary's Road and the south Perimeter Highway. It handles
20% to 25% of Winnipeg's sewage.*

*The problem relates to the third of four steps wastewater goes through
at the plant. After having grit, sediments and other large particles
removed, sewage is sent to a tank filled with what Shkolny called "a
bacterial cocktail" that sees beneficial bacteria consume the harmful
organics contained in the sewage.*

*To the city's surprise, all the bacteria in the tank have died, meaning
there is nothing alive in the tank to eat the harmful organics, so they're
being discharged into the river.*

*The water now flowing into the Red contains about double the
ammonia it did prior to the malfunction and fecal coliform counts —
most commonly the E. coli bacteria — that are not only well above
recreational guidelines, but amounts in violation of the city's
environmental licence.*

The city is licensed to keep the count below 200 bacteria per 100 ml of water, and current levels are between 1,000 and 2,000, city officials said.

The plant's managers still aren't sure what went wrong but have assembled a team of experts to try to solve the problem.

Shkolny said the city is only making the problem public now because it expected to be able to solve the issue earlier but has since realized its efforts were unsuccessful.

"Our hope would be to be see some improved quality soon and that in a month, maybe longer, we'll be back to normal," he said.

Appendix 16a

San Diego power outage caused 2 million gallons of raw sewage spilled into Los Penasquitos Lagoon and Sweetwater River
By chillymanjaro
– September 9, 2011
http://thewatchers.adorraeli.com/2011/09/09/san-diego-power-outage-caused-2-million-gallons-of-raw-sewage-spilled-into-los-penasquitos-lagoon-and-sweetwater-river/

The city of San Diego water system has lost power, causing two sewage pumps stations to spill raw sewage.

1.9 million gallons of raw sewage spilled into Los Penasquitos Lagoon according to Mark McPherson, the county's chief of land and water quality. Officials shut down about 10 miles of beaches north of Scripps Pier through Del Mar and Solana Beach. A second spill of 120,000 gallons into Sweetwater River, which flows into the southern part of San Diego Bay, closed Pepper Park and Bayside Park and the San Diego Bay area accessed from Silver Strand. Some of the pump stations don't have backup power, leading to an overflow of sewage during the outage.

Officials are requiring various neighborhoods throughout San Diego County to boil water prior to consumption to ensure its safety, the city's water department said in a statement Friday.

San Diego was urged to be cautious with water until further notice, and residents in Scripps-Miramar, Tierrasanta, San Carlos, Bernardo Hights, Scripps Ranch, La Jolla-Soledad, Otay Mesa, and the College grove areas are ordered to boil water prior to consumption.

Reduced water pressure throughout the San Diego County may allow for possible contamination, according to city officials. Defunct generators in the area had caused water pressure to significantly reduce, causing a high likelihood of backflow from homes and businesses. As water flows back into clean supply, the chance for contamination is high.

Cities throughout San Diego County are urged to reduce the use of water Friday and throughout the weekend, and to use boiled water, or opt for bottled or filtered options until further notice.

According to SDG&E, the power outage began with a major transmission outage in western Arizona that caused a loss of power to

*southern California. Shortly afterward, the San Onofre Generating
Station went off line. As a result, SDG&E did not have adequate
resources on its system to keep power on across its service territory.*

*SDG&E is a regulated public utility that provides energy service to 3.5
million consumers through 1.4 million electric meters and more than
850,000 natural gas meters in San Diego and southern Orange
counties. The utility's area spans 4,100 square miles*

Appendix 16b

Power outage closes beaches and weakens grid
Sept. 9, 2011 | KPCC & wires
http://www.scpr.org/news/2011/09/09/28741/power-restored-14-million-socal/

Power was restored sooner than expected for 1.4 million Southern California residents, who lost electricity in massive blackouts Thursday. Power officials are now saying sewage spills caused by the outage may have contaminated drinking water in high elevation areas. Residents of those areas are being warned to boil their water.

Most of the people affected were San Diego Gas & Electric customers, but as power outages swept Coachella Valley, southern Orange County and parts of Mexico, the number of residents without power rose to 5 million.

Mexico's electrical utility says the lights are now on for about 1.1 million customers, or about 97 percent of those who lost power. A few industrial clients are still without it.

Power has also been restored to all 56,000 customers in Yuma, Ariz.

Electricity is back in San Diego, but city schools, state universities and community colleges in the area are closed for the day.

The loss of electricity contributed to huge financial losses for grocers and restaurants that were unable to refrigerate their goods during the outage.

Brian McGray says he threw out $1,500 worth of steak, along with chicken, cheese and eggs at The Riders Club Cafe in San Clemente.

Darren Gorski buys supplies for The Fish Market in San Diego and spent the night watching thermometers on five insulated refrigerators full of seafood.

He says $50,000 in fish survived the outage, but that he lost thousands in dairy products.

The outage also cut power to sewage pumps in San Diego. More than 2 million gallons of raw sewage has spilled into the ocean and onto some beaches.

Mark McPherson is San Diego County's chief of land and water quality, he says the Department of Environmental Health has closed portions of the coast line due to massive sewage spills.

"One of them is 1.9 million gallons that went into Los Penasquitos Lagoon, which empties into the Pacific Ocean at Torrey Pines State Beach. And the Department of Environmental Health has issued closure for the beaches five miles south and north of that point where it comes into the ocean."

McPherson says a second spill let 120,000 gallons of sewage pour into the Sweetwater River, which empties into San Diego Bay. Two nearby parks have been closed.

No homes are currently affected by the spills but beach and park closures are expected to continue throughout the weekend.

Erin Coller, a spokeswoman for SDG&E, said the overall power system is still fragile, meaning customers should try and conserve energy.

"There was an issue on the interconnected grid with a major transmission outage in Western Arizona that caused a loss of power to Southern California," Coller said.

"Then shortly afterward the San Onofre Nuclear Generating Station went offline. So as a result we didn't have adequate resources on our system to keep the power on across our service territory."

The outage triggered a shutdown of San Onofre's two nuclear reactors. John Wayne Airport was not affected by the power outage, but the San Diego International Airport's air traffic control tower no longer had power and operations there were halted except for inbound flights, which stopped at 6:30 p.m.

Amtrak spokesman Marc Magliari says two Pacific Surfliner trains couldn't continue southward past Los Angeles after power was lost Thursday afternoon.

Officials at Phoenix-based Arizona Public Service Co. say the outage occurred after an electrical worker removed a piece of monitoring equipment at a substation.

According to the Arizona Daily Star, Mike Niggli, chief operating officer of SGD&E, said, "To my knowledge this is the first time we've lost an entire system."

It's still unclear how many power customers elsewhere are still without power.

Chapter 14 - Drinking Water Concerns

As cited, "contaminated water has always been an important agent in the spread of disease. Ingestion may cause gastrointestinal diseases, and skin diseases may be caused by immersion."[1]

"Water treatment and disinfection have markedly reduced the incidence of many diseases, but the need for constant vigilance and enforcement of standards is highlighted by the occasional water-borne disease epidemic."[2]

"Microbiological criteria are presently undergoing re-evaluation throughout the world, and the historical dependence upon total and fecal coliforms is being supplanted by more specific, epidemiologically-derived indicators of water quality."[3]

"Eventually, each use of the water and source of contamination should have a representative specific indicator associated with a regression equation so that disease risks can be quantified."[4]

"Once this occurs, the costs of achieving a given water quality could be directly compared with the savings in health care costs resulting from this level of water quality."[5]

"Organisms which are pathogenic to humans are rarely capable of survival as free-living organisms for very long."[6]

"Virtually all the water-borne diseases affecting humans are a result of poor waste treatment and disposal practices; the disposal of disease organisms carrying resistance to antibiotic drugs is of particular concern."[7]

"The harvesting of shellfish for human consumption is particularly sensitive to water quality since these organisms filter out and concentrate pathogens found in the water at relatively low levels. Contaminated shellfish have been responsible for many disease epidemics."[8]

"Since the direct monitoring of all possible pathogens would be too slow and uneconomical for routine water quality control, microbiological water quality is commonly estimated or monitored using a single or a few indicator organisms."[9]

"The validity and usefulness of the indicator concept depends upon the existence of a constant quantitative relationship between the indicator organism and the pathogens it is monitoring."[10]

"Establishing such relationships is a complex process, and several indicators are necessary if acceptable health risks are to be assigned to all pathogens in an economical manner."[11]

In other words, cost becomes a factor – unless "this indicator-to-pathogen ratio may change when an epidemic occurs or if changes take place in the quality of health care and treatment."[12]

There are other concerns about what appears in drinking water.

Zinc

As cited, "Zinc is commonly found in many natural waters."[1]

"The deterioration of galvanized iron and leaching of brass can add substantial amounts of zinc to water."[2]

"Industrial effluents may also contribute large amounts of zinc to drinking water."[3]

"Zinc is essential to human metabolism and has been found to be necessary for proper body growth."[4]

"Although essential in our diet, high zinc concentrations in water can irritate the human digestive system."[5]

"Levels above 5 mg/L cause a bitter metallic taste and opalescence in alkaline drinking water."[6]

"High concentrations of zinc suggest the presence of lead and cadmium, common impurities from the galvanizing process."[7]

"The EPA recommended limit is 5 mg/L."[8]

Manganese

"Manganese in water is a common, naturally occurring problem."[9]

"It can also be introduced by industry."[10]

"Manganese is usually found in combination with iron."[11]

"It causes a bitter taste in water, and at concentrations above 0.05 mg/L, it causes dark scale in pipes and water heaters."[12]

"High levels of manganese cause black staining of plumbing fixtures and laundry, and clogs up submersible pumps and pipes."[13]

"The EPA recommended limit is 0.05 mg/L."[14]

Lead

"Lead is a metal found in natural deposits as ores containing other elements."[15]

"It is sometimes used in household plumbing materials or in water service lines used to bring water from the main to the home."[16]

"The main source of lead in drinking water is from corrosion of household plumbing systems."[17]

"Lead and its compounds are poisonous and accumulate in the bone structure when ingested in amounts exceeding the natural elimination rate of about 300 ug per day."[18]

"Accumulation of significant amounts of lead in the body may cause severe and permanent brain damage, convulsions, and death."[19]

"The EPA recommended limit for lead is 0.015 mg/L."[20]

As further cited, "a 1986 EPA survey estimated that 40 million Americans (one in five) were using drinking water that contained potentially hazardous levels of lead."[21]

"This finding led to changes in the Safe Drinking Water Act to require the use of "lead-free" pipe, solder, and flux in the installation or repair of home and commercial plumbing connected to public water systems."[22]

"Acute lead poisoning can cause severe brain damage and death. The effects of chronic, low-level exposure, however, are more subtle. The developing nervous systems of fetuses, infants, and children are particularly vulnerable. Recent studies show that lead exposure at a young age can cause permanent learning disabilities and hyperactive behavior. Low-level lead exposure also is associated with elevated blood pressure, chronic anemia, and peripheral nerve damage."[23]

"Natural water usually contains very little lead. Contamination generally occurs in the water distribution system or in the pipes of a home or facility. Lead pipes, brass faucets and lead solder used to join copper pipes are the culprits. If your home was built before 1986 when the nationwide ban on lead pipes and lead solder went into effect, it is likely to have lead-soldered plumbing."[24]

"The severity of lead contamination depends in part on how "corrosive" your water is. Soft or acidic water is more likely to corrode plumbing and fixtures, leaching out lead. According to the EPA, about 80 percent of public water utilities deliver water that is moderately or highly corrosive."[25]

"Techniques such as adding lime (calcium oxide) to reduce water acidity can greatly reduce lead levels at the tap."[26]

"Consumers can follow a number of simple practices to help reduce the level of lead at the tap:

1. Cook with and drink only cold water. Hot water tends to dissolve more lead from pipes.

2. Don't drink the first water out of your tap in the morning. Let the water run for about one minute until a change in temperature occurs.

3. For private wells, consider water treatment devices such as calcite filters that reduce acidity and make water less corrosive. Certain point-of-purchase treatment devices (e.g., some ion-exchange filters, reverse osmosis devices and distillation units) also can remove lead.

4. If lead levels remain high, consider bottled water for drinking and cooking purposes."[27]

"Everyone is exposed to low levels of lead through food, drinking water, air, household dust, soil, and some consumer products."[28]

"However, ongoing exposure to even small amounts of lead may be harmful to your health."[29]

"The amount of lead in the environment increased during the industrial revolution, and again significantly in the 1920s with the introduction of leaded gasoline."[30]

"While lead is not deliberately added to foods, low levels of lead have been detected in a variety of foods. Lead is introduced to foods through uptake from soil into plants and by airborne lead falling onto plant surfaces."[31]

"Lead is released into air through industrial emissions, smelters and refineries."[32]

"Dust and soil can be significant sources of exposure to lead for toddlers. Lead levels in soil tend to be higher in cities, near roadways, and around industrial sources that use or emit lead, near weapon firing ranges, or next to buildings where crumbling leaded paint has fallen into the soil."[33]

"The EPA currently requires that public utilities ensure that lead levels at the customer's tap not exceed 15 parts per billion (ppb)."[34]

Sulfate

As reported, "Sulfates occur naturally in groundwater combined with calcium, magnesium and sodium as sulfate salts."[35]

"Sulfate content in excess of 250 to 500 ppm (mg/l) may give water a bitter taste and have a laxative effect on individuals not adapted to the water."[36]

"Water that smells like rotten eggs has a high level of hydrogen sulfide gas."[37]

"The gas may occur naturally in water near oil or gas fields or as the result of bacterial contamination."[38]

Tannins

"Tannins are organic materials dissolved in the water."[39]

"It is a product of decomposed plant material which occurs in natural waters."[40]

"They can interfere with water softener filter resin beds and impart a distinctive yellow-brown color to the water."[41]

"Tannins can also be associated with organically bound iron."[42]

"Levels above 0.5 mg/L cause light brown or yellowish stains on laundry and fixtures."[43]

"These levels can also affect the taste of foods and beverages."[44]

Footnotes

1 – 8. *BC Drinking Water*
http://www.env.gov.bc.ca/wat/wq/BCguidelines/microbiology/microbiology.html

9 – 14. *Water Test, Inc. tests for the following contaminants*:
http://www.watertestinc.com/contaminants.html

15 – 20. *Water Test, Inc. tests for the following contaminants*:
http://www.watertestinc.com/contaminants.html

21 - 30. *Drinking water quality and health*
by P. Kendall1(3/2010)
http://www.ext.colostate.edu/pubs/foodnut/09307.html

31 - 33. *Lead*

http://www.hc-sc.gc.ca/hl-vs/iyh-vsv/environ/lead-plomb-eng.php

34. *Water Test, Inc. tests for the following contaminants*:
http://www.watertestinc.com/contaminants.html

35 - 38. *Drinking water quality and health*
by P. Kendall1(3/2010)
http://www.ext.colostate.edu/pubs/foodnut/09307.html

39 – 44. *Water Test, Inc. tests for the following contaminants*:
http://www.watertestinc.com/contaminants.html

Chapter 15 - More Drinking Water Concerns

"According to a 1999 study by the National Academy of Sciences, arsenic in drinking water causes bladder, lung and skin cancer, and may cause kidney and liver cancer. The study also found that arsenic harms the central and peripheral nervous systems, as well as heart and blood vessels, and causes serious skin problems. It also may cause birth defects and reproductive problems."[1]

"In a February 2000 report, NRDC analyzed data compiled by the U.S. Environmental Protection Agency on arsenic in drinking water in 25 states. Our most conservative estimates based on the data indicated that more than 34 million Americans were drinking tap water supplied by systems containing average levels of arsenic that posed unacceptable cancer risks. We consider it likely that as many as 56 million people in those 25 states were drinking water with arsenic at unsafe levels -- and that's just the 25 states that reported arsenic information to the EPA."[2]

In addition, it was reported "arsenic may enter lakes, rivers or underground water naturally, when mineral deposits or rocks containing arsenic dissolve. Arsenic may also get into water through the discharge of industrial wastes and by the deposit of arsenic particles in dust, or dissolved in rain or snow."[3]

"These arsenic particles can enter the environment through:

- the burning of fossil fuels (especially coal);

- metal production (such as gold and base metal mining);

- agricultural use (in pesticides and feed additives); or

- waste burning."[4]

"Arsenic in drinking water is absorbed by the body when you swallow it, and distributed by the bloodstream. It does not enter the body through the skin or by inhalation during bathing or showering. The highest levels of arsenic are found in nails and hair, which accumulate arsenic over time. Your body gets rid of arsenic mostly through urine, with smaller amounts removed through the skin, hair, nails and sweat."[5]

"Health Canada and the International Agency for Research on Cancer consider arsenic a human cancer-causing agent. Its effects have been studied in a population in Taiwan where the drinking water contains naturally high levels of arsenic (over 0.35 ppm)."[6]

"Arsenic is one of the many chemicals for which Health Canada has set guidelines. A new guideline has been established at 0.010 milligrams per litre, and will continue to be reviewed to reflect new treatment methods and new information on health risks as they become available."[7]

"Data collected indicate that the levels of arsenic in Canadian drinking water are generally less than 0.005 milligrams per litre (0.005 parts per million - ppm), although concentrations may be higher in some areas."[8]

Giardia

"Chlorination and filtration are effective controls for most bacteria."[9]

"However, a tiny one-celled parasite not readily killed by chlorination, Giardia lamblia, deserves special discussion."[10]

"Giardia has become an increasingly common problem in rural and mountain communities with inadequate filtration systems."[11]

"Giardia is mostly found in surface waters such as mountain streams and lakes, not groundwater. Because one cannot see, taste, or smell giardia, it is best not to drink water directly from mountain streams or lakes."[12]

"Once ingested, the giardia cyst develops into a trophozoite that attaches to the wall of the small intestine. Disease symptoms usually include diarrhea with cramping and gas, dehydration, weakness and loss of appetite. Symptoms may take seven to 10 days to appear and last up to six weeks. Most people are unaware at the time of ingestion that they have been infected."[13]

"Laboratory identification can confirm the disease by diagnosis of the organism in the stool. The disease is curable with prescribed medication. If untreated, the symptoms may disappear on their own and reoccur intermittently over a period of months."[14]

"Dogs, like people, can get infected with giardia. Unless carefully controlled, dogs can contaminate the water and continue the chain of infection from animals to humans."[15]

Cryptosporidium

As further note, "Giardia and Cryptosporidium are microscopic parasites that can be found in water. Giardia causes an intestinal illness called giardiasis or 'beaver fever.'"[16]

"Cryptosporidium is responsible for a similar illness called cryptosporidiosis."[17]

"Both parasites produce cysts that are very resistant to harsh environmental conditions. When ingested, they germinate, reproduce, and cause illness. After feeding, the parasites form new cysts, which are then passed in the faeces. Studies with human volunteers have shown that ingestion of only a few cysts will cause illness."[18]

"Diarrhoea, abdominal cramps, gas, malaise, and weight loss are the most common symptoms caused by Giardia. Vomiting, chills, headache, and fever may also occur. These symptoms usually surface six to 16 days after the initial contact and can continue as long as one month."[19]

"The symptoms of cryptosporidiosis are similar; the most common include watery diarrhoea, abdominal cramps, nausea, and headaches. These symptoms occur within two to 25 days of infection and usually last one or two weeks; in some cases they stick around for up to a month."[20]

Pseudomonas aeruginosa

Another concern reported, "Pseudomonas aeruginosa is a common opportunistic pathogen of man which causes ear infections and other non-gastrointestinal infections when present in recreational waters."[21]

"These include hot tubs, whirl pools and swimming pools."[22]

"It is not adequately monitored by such gastrointestinal indicators as the coliforms and needs to be monitored directly."[23]

"Enterococci are better indicators than fecal coliforms and most closely approach the ideal characteristics of an indicator for gastrointestinal diseases."[24]

"They are the best indicators for recreational uses, especially in marine waters where E. coli do not survive as well."[25]

"They are not as useful in water with high organic wastes from vegetable processing, but are good in assessing reservoir quality, sewage-contaminated water supplies and chlorinated water high in organics."[26]

"Escherichia coli is a better indicator than fecal coliforms since Klebsiella is not enumerated in the E. coli test."[27]

"Klebsiella may multiply in water containing pulp mill effluent and other organics, which is not contaminated by human sewage, and thus gives false positives for fecal contamination."[28]

"These indicators have higher correlations with specific types of disease under specified conditions than does the fecal coliform test, and are recommended as replacements for fecal coliforms."[29]

Perchlorate

"Perchlorate is chemical used in military and industrial products such as solid rocket fuels, munitions, explosives and fireworks, road flares and air bag inflation systems. It can also occur naturally at low levels in the environment, for example in certain arid regions of the world."[30]

"To date, perchlorate has only been found in one Canadian public drinking water supply, at a concentration below 1 part per billion (ppb); a ppb is equivalent to the content of half a teaspoon in an olympic-sized swimming pool."[31]

However, later in this citation, it mentions "where a contamination of drinking water supplies in Canada has occurred, Health Canada recommends a drinking water guidance value of 6 ppb, based on a review of existing health risk assessments from other agencies."[32]

These two previous statements are puzzling, in the first it says "perchlorate has only been found in one Canadian public drinking water supply" yet later in this citation it says ""where a contamination of drinking water supplies in Canada has occurred" which seems to be more than one contamination?

However, continuing, "American data show that perchlorate may also be found in foods such as lettuce, milk and bottled water."[33]

"Perchlorate inhibits the transfer of iodide from the blood to the thyroid gland, which is required for the gland to produce hormones essential for metabolism and growth. Although short-term fluctuations in thyroid hormones are not a concern to healthy adults, long-term disruptions may result in hypothyroidism and related changes in metabolism, decreased mental performance, and altered development."[34]

"Scientific studies and guideline development related to perchlorate are on-going. In 2005, the American National Academy of Science produced a report on the health implications of perchlorate ingestion, which has been used by the U.S. Environmental Protection Agency to establish a preliminary clean-up goal of 24.5 ppb for perchlorate in water. There is currently no enforceable national drinking water standard for perchlorate either in Canada or in the United States, although various states have implemented guidelines or goals ranging from 1 ppb to 18 ppb for perchlorate in drinking water."[35]

"Technologies for reducing perchlorate in the water supply are available. There are municipal-scale treatment systems and several certified reverse osmosis residential systems that reduce perchlorate to 6 ppb or less."[36]

Fluoride

As reported, "most Canadians are exposed to fluorides on a daily basis, through the trace amounts that are found in almost all foods and through those that are added to some drinking water supplies to prevent tooth decay."[37]

"The use of fluoride for the prevention of dental cavities is endorsed by over 90 national and international professional health organizations including Health Canada, the Canadian Public Health Association, the Canadian Dental Association, the Canadian Medical Association, the US Food and Drug Administration and the World Health Organization."[38]

"Fluorides protect tooth enamel against the acids that cause tooth decay. Many studies have shown that fluoridated drinking water significantly reduces the number of cavities in children's teeth. Fluoride is used in many communities across Canada, spanning most provinces and territories. About 45 per cent of Canadians receive fluoridated water."[39]

With respect to children, "as young children tend to swallow toothpaste when they are brushing, the following guidelines have been established to balance their risk of developing dental fluorosis with the dental health benefits of fluoride:

- *Children up to 3 years of age should have their teeth and gums brushed by an adult. Parents should consult a health professional to determine whether their child under 36 months of age is at risk of developing tooth decay. If the child is at risk of developing tooth decay, then they should have their teeth brushed by an adult using a minimal amount (rice sized grain) of fluoridated toothpaste. It has been determined that use of fluoride toothpaste in a small amount effectively balances between the benefit of fluoride and the risk of developing fluorosis. If the child is not considered at risk, it is recommended their teeth be brushed by an adult using a toothbrush moistened only with water.*

- *Children 3 - 6 years of age should be assisted with brushing their teeth by an adult and use only a small amount (i.e., green pea-sized portion) of fluoridated toothpaste."[40]*

With respect to adults, "high levels of fluorides consumed for a very long period of time may lead to skeletal fluorosis. These levels are much higher than those to which the average Canadian is exposed daily. Skeletal fluorosis is a progressive but not life-threatening disease in which bones increase in density and become more brittle. In mild cases, the symptoms may include pain and stiff joints. In more severe cases, the symptoms may include difficulty in moving, deformed bones and a greater risk of bone fractures."[41]

"Like most elements, fluoride appears to be both beneficial to health and potentially toxic."[42]

"The goal is to determine the optimum level and then decide how best to achieve that level."[43]

"The EPA currently sets the maximum allowable level of sodium fluoride in drinking water (natural or added) at 4 milligrams per liter (4 parts per million) and the maximum recommended level at 2 milligrams per liter."[44]

"Health Canada has established the guideline for fluoride in drinking water as a maximum acceptable concentration of 1.5 parts per million with water "at, or below, this maximum acceptable concentration does not pose a risk to human health"[45]

According to a press release from the City of Winnipeg date March 16th, 2011 they have lowered the fluoride level in drinking water from the original 0.85 ppm to 0.7 ppm "in response to new provincial guideline" that was issued on March 15th of the same year. This change was "consistent with a review from Health Canada," and as such Manitoba Health has made such changes to "public drinking water supplies to prevent dental caries"[46]

Footnotes

1 – 2. *Arsenic in Drinking Water*
Natural Resources Defense Council
http://www.nrdc.org/water/drinking/qarsenic.asp

3 - 8. *Arsenic*
http://www.hc-sc.gc.ca/hl-vs/iyh-vsv/environ/arsenic-eng.php

9 - 15. *Drinking water quality and health*
by P. Kendall1(3/2010)
http://www.ext.colostate.edu/pubs/foodnut/09307.html

16 - 20. *Giardia and Cryptosporidium*
http://www.hc-sc.gc.ca/ewh-semt/pubs/water-eau/giardia_cryptosporidium-eng.php

21 - 29. *BC Drinking Water*
http://www.env.gov.bc.ca/wat/wq/BCguidelines/microbiology/microbiology.html

30 - 36. *Perchlorate - MILITARY*
http://www.hc-sc.gc.ca/ewh-semt/pubs/water-eau/perchlorate-eng.php

37 - 41. *Fluoride and Human Health*
http://www.hc-sc.gc.ca/hl-vs/iyh-vsv/environ/fluor-eng.php

42 - 44. *Drinking water quality and health*
by P. Kendall1(3/2010)
http://www.ext.colostate.edu/pubs/foodnut/09307.html

45 - 46. *Fluoride*
http://www2.fluoridealert.org/Alert/Canada/Manitoba/Winnipeg-Fluoride-level-in-drinking-water-lowered-to-0.7ppm-from-0.85ppm-in-response-to-new-provincial-guideline

Also see: *City lowers fluoride in drinking water*
CBC News
Posted: Mar 16, 2011
http://www.cbc.ca/news/canada/manitoba/story/2011/03/16/man-flouride-drinking-water-winnipeg.html

Chapter 16 - Organic & Inorganic Minerals

As cited, "most knowledgeable people today recognize that the body must have certain minerals to accomplish its work and preserve its health. However, only a few realize that these minerals must be in their organic state to do us any good at all."[1]

"Minerals are inorganic as they exist naturally in the soil and water."[2]

"Minerals are organic as they exist in plants and animals."[3]

"Only plants can transform inorganic minerals into organic minerals."[4]

"Animals must eat plants or plant-eating animals to obtain their organic minerals."[5]

"Inorganic minerals are useless and injurious to the animal organism."[6]

"Because inorganic minerals and organic minerals have the same chemical compositions, they were confused by the early nutritionists."[7]

"The mineral, iron, in the bloodstream has the same chemical composition as the mineral, iron, in a nail—iron is iron, after all."[8]

"However, these nutritionists incorrectly reasoned that there were no other differences between these two forms of iron. As a consequence, there actually were iron mineral supplements that consisted of surplus powdered nails."[9]

"These nutritionists made an error in reasoning by assuming that a chemical similarity in minerals also meant there was a nutritive similarity between organic and inorganic minerals. While it is true that the same minerals found in the human body are also found in the soil and water it is wrong to assume that the minerals in the soil are food for man. We are not soil eaters—we are plant eaters."[10]

As well, "chemically, it is true that iron in the bloodstream and iron in nails are the same and that calcium in rocks (known as dolomite) is identical to calcium in the bones."[11]

"However, it is a grave error to believe that the body can digest and assimilate and utilize powdered nails and crushed rocks."[12]

As further explained, "Organic compounds contain carbon-hydrogen bonds. Inorganic compounds don't."[13]

This is a much better definition, allowing us to call sodium acetylide "organic" [see **Figure 16**, appearing below] but calcium carbide "inorganic" [see **Figure 17**, appearing below], but it doesn't always work.[14]

Figure 16: sodium acetylide "organic"

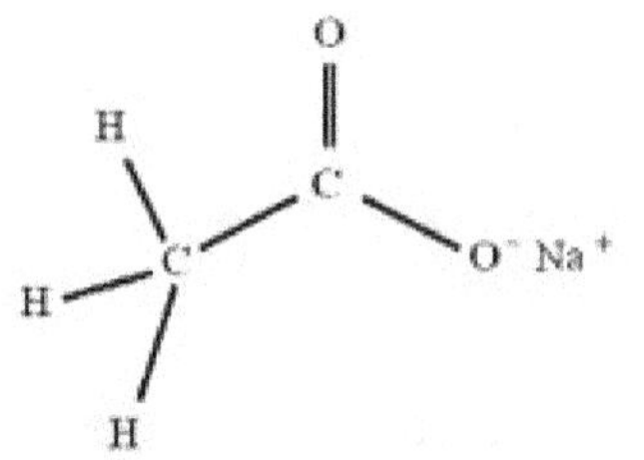

Reference to: *What is the difference between an inorganic and organic compound?*
Date: Thu Nov 30 12:02:32 2000
Posted By: Dan Berger, Faculty Chemistry/Science, Bluffton College
Area of science: Chemistry
ID: 975377363.Ch
http://www.madsci.org/posts/archives/2000-12/975719013.Ch.r.html

---

Figure 17: calcium carbide "inorganic"

Reference to: *Calcium Carbide Formula*
http://www.ask.com/questions-about/Calcium-Carbide-Formula

This is a much better definition, allowing us to call sodium acetylide "organic" but calcium carbide "inorganic, but it doesn't always work; but what about tetracyanoethylene [see **Figure 18**, appearing below], which is indubitably organic but contains no hydrogen atoms at all?"[14]

Figure 18: tetracyanoethylene

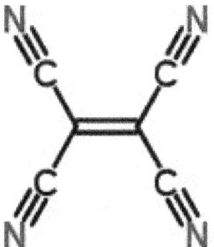

Reference to: *What is the difference between an inorganic and organic compound?*
Date: Thu Nov 30 12:02:32 2000
Posted By: Dan Berger, Faculty Chemistry/Science, Bluffton College
Area of science: Chemistry
ID: 975377363.Ch
http://www.madsci.org/posts/archives/2000-12/975719013.Ch.r.html

"Nevertheless, almost all 'organic carbon' has carbon-hydrogen bonds, while all "inorganic carbon" does not have carbon-hydrogen bonds. Even if you call e.g. sodium acetate inorganic, everyone will agree that the carbon atom which is attached to the hydrogens is 'organic carbon'."[15]

Footnotes

1 - 12. *Organic And Inorganic Minerals*
http://www.rawfoodexplained.com/minerals/organic-and-inorganic-minerals.html

13 - 15. *What is the difference between an inorganic and organic compound?*
Date: Thu Nov 30 12:02:32 2000
Posted By: Dan Berger, Faculty Chemistry/Science, Bluffton College
Area of science: Chemistry
ID: 975377363.Ch
http://www.madsci.org/posts/archives/2000-12/975719013.Ch.r.html

Chapter 17 - More about Minerals

As cited, "the term 'organic chemical' includes such products as pesticides, herbicides, petroleum products and industrial solvents."[1]

"Hundreds of different organic chemicals have been found in drinking water from accidental spills, improper disposal or non-point movement through soils to groundwater."[2]

"Today, municipalities are required to monitor more than 50 organic chemicals under the Safe Drinking Water Act."[3]

"As with other contaminants, the danger from organic chemicals in water is hard to assess."[4]

"In high doses these chemicals may cause various problems including increased risk of cancer, impaired nervous system or damage to the heart."[5]

"In low doses, organic chemicals may have cumulative effects, but less is known about their nature or magnitude."[6]

A list of Microorganisms is cited in **Appendix 17a**, cited below.[7]

A list of Disinfectants is cited in **Appendix 17b**, cited below.[8]

A list of Disinfection By-products is cited in **Appendix 17c**, cited below.[9]

A list of Inorganic Chemicals is cited in **Appendix 17d**, cited below.[10]

A list of Organic Chemicals is cited in **Appendix 17e**, cited below.[11]

Footnotes

1 - 6. *Drinking water quality and health*
By P. Kendall1(3/2010)
http://www.ext.colostate.edu/pubs/foodnut/09307.html

7 - 11. Drinking Water Contaminants
http://water.epa.gov/drink/contaminants/index.cfm

Appendix 17a

A list of Microorganisms

Microorganisms

Contaminant	MCLG[1] (mg/L)[2]	MCL or TT[1] (mg/L)[2]	Potential Health Effects from Long-Term Exposure Above the MCL (unless specified as short-term)
Cryptosporidium	zero	TT [3]	Gastrointestinal illness (e.g., diarrhea, vomiting, cramps)
Giardia lamblia	zero	TT[3]	Gastrointestinal illness (e.g., diarrhea, vomiting, cramps)
Heterotrophic plate count	n/a	TT[3]	HPC has no health effects; it is an analytic method used to measure the variety of bacteria that are common in water. The lower the concentration of bacteria in drinking water, the better maintained the water system is.
Legionella	zero	TT[3]	Legionnaire's Disease, a type of pneumonia
Total Coliforms (including fecal coliform and E. Coli)	zero	5.0%[4]	Not a health threat in itself; it is used to indicate whether other potentially harmful bacteria may be present[5]
Turbidity	n/a	TT[3]	Turbidity is a measure of the cloudiness of water. It is used to indicate water quality and filtration effectiveness (e.g., whether disease-causing organisms are present). Higher turbidity levels are often associated with higher levels of disease-causing microorganisms such as viruses, parasites and some bacteria. These organisms can cause symptoms such as nausea, cramps, diarrhea, and associated headaches.
Viruses (enteric)	zero	TT[3]	Gastrointestinal illness (e.g., diarrhea, vomiting, cramps)

Cited from: *Drinking Water Contaminants*
http://water.epa.gov/drink/contaminants/index.cfm

Appendix 17b

A list of Disinfectants

Disinfectants

Contaminant	MCLG[1] (mg/L)[2]	MCL or TT[1] (mg/L)[2]	Potential Health Effects from Long-Term Exposure Above the MCL (unless specified as short-term)
Chloramines (as Cl_2)	MRDLG=4[1]	MRDL=4.0[1]	Eye/nose irritation; stomach discomfort, anemia
Chlorine (as Cl_2)	MRDLG=4[1]	MRDL=4.0[1]	Eye/nose irritation; stomach discomfort
Chlorine dioxide (as ClO_2)	MRDLG=0.8[1]	MRDL=0.8[1]	Anemia; infants & young children: nervous system effects

Cited from: *Drinking Water Contaminants*
http://water.epa.gov/drink/contaminants/index.cfm

Appendix 17c

A list of Disinfection By-products

Disinfection Byproducts

Contaminant	MCLG[1] (mg/L)[2]	MCL or TT[1] (mg/L)[2]	Potential Health Effects from Long-Term Exposure Above the MCL (unless specified as short-term)
Bromate	zero	0.010	Increased risk of cancer
Chlorite	0.8	1.0	Anemia; infants & young children: nervous system effects
Haloacetic acids (HAA5)	n/a[6]	0.060[7]	Increased risk of cancer
Total Trihalomethanes (TTHMs)	--> n/a[6]	--> 0.080[7]	Liver, kidney or central nervous system problems; increased risk of cancer

Cited from: *Drinking Water Contaminants*
http://water.epa.gov/drink/contaminants/index.cfm

Appendix 17d

A list of Inorganic Chemicals

Inorganic Chemicals

Contaminant	MCLG[1] (mg/L)[2]	MCL or TT[1] (mg/L)[2]	Potential Health Effects from Long-Term Exposure Above the MCL (unless specified as short-term)
Antimony	0.006	0.006	Increase in blood cholesterol; decrease in blood sugar
Arsenic	0[2]	0.010 as of 01/23/06	Skin damage or problems with circulatory systems, and may have increased risk of getting cancer
Asbestos (fiber >10 micrometers)	7 million fibers per liter	7 MFL	Increased risk of developing benign intestinal polyps
Barium	2	2	Increase in blood pressure
Beryllium	0.004	0.004	Intestinal lesions
Cadmium	0.005	0.005	Kidney damage
Chromium (total)	0.1	0.1	Allergic dermatitis
Copper	1.3	TT[2]; Action Level=1.3	Short term exposure: Gastrointestinal distress. Long term exposure: Liver or kidney damage. People with Wilson's Disease should consult their personal doctor if the amount of copper in their water exceeds the action level

Cited from: *Drinking Water Contaminants*

http://water.epa.gov/drink/contaminants/index.cfm

Appendix 17d continued

A list of Inorganic Chemicals

Inorganic Chemicals

Contaminant	MCLG[1] (mg/L)[2]	MCL or TT[1] (mg/L)[2]	Potential Health Effects from Long-Term Exposure Above the MCL (unless specified as short-term)
Cyanide (as free cyanide)	0.2	0.2	Nerve damage or thyroid problems
Fluoride	4.0	4.0	Bone disease (pain and tenderness of the bones); Children may get mottled teeth
Lead	zero	TT[2]; Action Level=0.015	Infants and children: Delays in physical or mental development; children could show slight deficits in attention span and learning abilities Adults: Kidney problems; high blood pressure
Mercury (inorganic)	0.002	0.002	Kidney damage
Nitrate (measured as Nitrogen)	10	10	Infants below the age of six months who drink water containing nitrate in excess of the MCL could become seriously ill and, if untreated, may die. Symptoms include shortness of breath and blue-baby syndrome.

Cited from: *Drinking Water Contaminants*
http://water.epa.gov/drink/contaminants/index.cfm

Appendix 17d continued

A list of Inorganic Chemicals

Contaminant	MCLG[1] (mg/L)[2]	MCL or TT[1] (mg/L)[2]	Potential Health Effects from Long-Term Exposure Above the MCL (unless specified as short-term)
Nitrite (measured as Nitrogen)	1	1	Infants below the age of six months who drink water containing nitrite in excess of the MCL could become seriously ill and, if untreated, may die. Symptoms include shortness of breath and blue-baby syndrome.
Selenium	0.05	0.05	Hair or fingernail loss; numbness in fingers or toes; circulatory problems
Thallium	0.0005	0.002	Hair loss; changes in blood; kidney, intestine, or liver problems

Cited from: *Drinking Water Contaminants*
http://water.epa.gov/drink/contaminants/index.cfm

Appendix 17e

A list of Organic Chemicals

Organic Chemicals

Contaminant	MCLG[1] (mg/L)[2]	MCL or TT[1] (mg/L)[2]	Potential Health Effects from Long-Term Exposure Above the MCL (unless specified as short-term)
Acrylamide	zero	TT[9]	Nervous system or blood problems; increased risk of cancer
Alachlor	zero	0.002	Eye, liver, kidney or spleen problems; anemia; increased risk of cancer
Atrazine	0.003	0.003	Cardiovascular system or reproductive problems
Benzene	zero	0.005	Anemia; decrease in blood platelets; increased risk of cancer
Benzo(a)pyrene (PAHs)	zero	0.0002	Reproductive difficulties; increased risk of cancer
Carbofuran	0.04	0.04	Problems with blood, nervous system, or reproductive system
Carbon tetrachloride	zero	0.005	Liver problems; increased risk of cancer
Chlordane	zero	0.002	Liver or nervous system problems; increased risk of cancer
Chlorobenzene	0.1	0.1	Liver or kidney problems
2,4-D	0.07	0.07	Kidney, liver, or adrenal gland problems

Cited from: *Drinking Water Contaminants*

http://water.epa.gov/drink/contaminants/index.cfm

Appendix 17e continued

A list of Organic Chemicals

Organic Chemicals

Contaminant	MCLG[1] (mg/L)[2]	MCL or TT[1] (mg/L)[2]	Potential Health Effects from Long-Term Exposure Above the MCL (unless specified as short-term)
Dalapon	0.2	0.2	Minor kidney changes
1,2-Dibromo-3-chloropropane (DBCP)	zero	0.0002	Reproductive difficulties; increased risk of cancer
o-Dichlorobenzene	0.6	0.6	Liver, kidney, or circulatory system problems
p-Dichlorobenzene	0.075	0.075	Anemia; liver, kidney or spleen damage; changes in blood
1,2-Dichloroethane	zero	0.005	Increased risk of cancer
1,1-Dichloroethylene	0.007	0.007	Liver problems
cis-1,2-Dichloroethylene	0.07	0.07	Liver problems
trans-1,2-Dichloroethylene	0.1	0.1	Liver problems

Cited from: *Drinking Water Contaminants*
http://water.epa.gov/drink/contaminants/index.cfm

Appendix 17e continued

A list of Organic Chemicals

Contaminant	MCLG[1] (mg/L)[2]	MCL or TT[1] (mg/L)[2]	Potential Health Effects from Long-Term Exposure Above the MCL (unless specified as short-term)
Dichloromethane	zero	0.005	Liver problems; increased risk of cancer
1,2-Dichloropropane	zero	0.005	Increased risk of cancer
Di(2-ethylhexyl) adipate	0.4	0.4	Weight loss, liver problems, or possible reproductive difficulties.
Di(2-ethylhexyl) phthalate	zero	0.006	Reproductive difficulties; liver problems; increased risk of cancer
Dinoseb	0.007	0.007	Reproductive difficulties
Dioxin (2,3,7,8-TCDD)	zero	0.00000003	Reproductive difficulties; increased risk of cancer
Diquat	0.02	0.02	Cataracts
Endothall	0.1	0.1	Stomach and intestinal problems
Endrin	0.002	0.002	Liver problems
Epichlorohydrin	zero	TT[9]	Increased cancer risk, and over a long period of time, stomach problems
Ethylbenzene	0.7	0.7	Liver or kidneys problems
Ethylene dibromide	zero	0.00005	Problems with liver, stomach, reproductive system, or kidneys; increased risk of cancer

Appendix 17e continued

A list of Organic Chemicals

Contaminant	MCLG[1] (mg/L)[2]	MCL or TT[1] (mg/L)[2]	Potential Health Effects from Long-Term Exposure Above the MCL (unless specified as short-term)
Glyphosate	0.7	0.7	Kidney problems; reproductive difficulties
Heptachlor	zero	0.0004	Liver damage; increased risk of cancer
Heptachlor epoxide	zero	0.0002	Liver damage; increased risk of cancer
Hexachlorobenzene	zero	0.001	Liver or kidney problems; reproductive difficulties; increased risk of cancer
Hexachlorocyclopentadiene	0.05	0.05	Kidney or stomach problems
Lindane	0.0002	0.0002	Liver or kidney problems
Methoxychlor	0.04	0.04	Reproductive difficulties
Oxamyl (Vydate)	0.2	0.2	Slight nervous system effects

Appendix 17e continued

A list of Organic Chemicals

Contaminant	MCLG[1] (mg/L)[2]	MCL or TT[1] (mg/L)[2]	Potential Health Effects from Long-Term Exposure Above the MCL (unless specified as short-term)
Pentachlorophenol	zero	0.001	Liver or kidney problems; increased cancer risk
Picloram	0.5	0.5	Liver problems
Simazine	0.004	0.004	Problems with blood
Styrene	0.1	0.1	Liver, kidney, or circulatory system problems
Tetrachloroethylene	zero	0.005	Liver problems; increased risk of cancer
Toluene	1	1	Nervous system, kidney, or liver problems
Toxaphene	zero	0.003	Kidney, liver, or thyroid problems; increased risk of cancer
2,4,5-TP (Silvex)	0.05	0.05	Liver problems

Cited from: *Drinking Water Contaminants*
http://water.epa.gov/drink/contaminants/index.cfm

Appendix 17e continued

A list of Organic Chemicals

Contaminant	MCLG[1] (mg/L)[2]	MCL or TT[1] (mg/L)[2]	Potential Health Effects from Long-Term Exposure Above the MCL (unless specified as short-term)
1,2,4-Trichlorobenzene	0.07	0.07	Changes in adrenal glands
1,1,1-Trichloroethane	0.20	0.2	Liver, nervous system, or circulatory problems
1,1,2-Trichloroethane	0.003	0.005	Liver, kidney, or immune system problems
Trichloroethylene	zero	0.005	Liver problems; increased risk of cancer
Vinyl chloride	zero	0.002	Increased risk of cancer
Xylenes (total)	10	10	Nervous system damage

Cited from: *Drinking Water Contaminants*
http://water.epa.gov/drink/contaminants/index.cfm

Chapter 18 - Even more concerns - Radon

"Radon is a radioactive gas, a decay product of uranium, that can dissolve into water supplies. The gas also is found in rocks and soils that contain granite, shale, phosphate, and pitchblende. It is odorless, colorless and tasteless."[1]

As further noted, "Radon is radioactive, which means that it breaks down - or 'decays' - to form other elements."[2]

"The rate of radon's radioactive decay is defined by its half-life, which is the time required for one half of any amount of the element to break down."[3]

"The half-life of radon is 3.8 days (Hem, 1985)."[4]

"Radon moves from its source in rocks and soils through voids and fractures. It can enter buildings as a gas through foundation cracks or dissolve in the ground water and be carried to water-supply wells."[5]

"The amount of radon in air or water commonly is reported in terms of activity with units of picocuries per liter of air or water."[6]

"An activity of 1 pCi/L (picocuries per liter) is about equal to the decay of two atoms of radon per minute in each liter of air or water (Otton, 1992)."[7]

For example, "ground-water samples collected from 267 wells were analyzed for radon as part of a water-quality reconnaissance of subunits of the Lower Susquehanna and Potomac River Basins conducted by the United States Geological Survey (USGS) as part of the National Water-Quality Assessment (NAWQA) program."[8]

"Exposure to radon has been recognized as a health risk, primarily as a cause of lung cancer. A study of miners found that inhaling the decay products of radon increases the chances of lung cancer (Robillard and others, 1991)."[9]

"Airborne radon has been cited by the Surgeon General of the United States as the second-leading cause of lung cancer and the United States Environmental Protection Agency (USEPA) has identified ground-water supplies as possible contributing sources of indoor radon."[10]

"Eighty percent of ground-water samples collected for this study were found to contain radon at activities greater than 300 pCi/L (picocuries per liter), the USEPA's proposed Maximum Contaminant Level for radon in drinking water, and 31 percent of samples contained radon at activities greater than 1,000 pCi/L."[11]

In addition, "Radon is most likely to be present in water from private wells or from small community systems. Large systems usually provide some kind of water treatment that aerates the water and disperses any radon gas that may be present."[12]

"Test results are expressed in picocuries of radon per liter of water (pCi/l)."[13]

"In general 10,000 pCi/l of radon in water contributes roughly 1 pCi/l of airborne radon throughout the house."[14]

"For waterborne radon, a simple step is to make sure your bathroom, laundry and kitchen are well ventilated. At moderate levels, this may adequately reduce your exposure to waterborne radon. However if you use a private well that has high levels of radon, water treatment devices such as granular activated carbon units and home aerators may be warranted."[15]

"EPA currently advises consumers to take action at total household air levels of 4 pCi/l."[16]

Footnotes

1. *Drinking water quality and health*
by P. Kendall1(3/2010)
http://www.ext.colostate.edu/pubs/foodnut/09307.html

2 - 11. Lindsey, Bruce D., and Ator, Scott W., 1996, *Radon in ground water of the Lower Susquehanna and Potomac River Basins:*
U.S. Geological Survey Water-Resources Investigations Report 96-4156
pa.water.usgs.gov/reports/wrir_96-4156/report.html
U.S. Geological Survey
Branch of Information Services
Box 25286, Denver, Colorado 80225-0286
http://pa.water.usgs.gov/reports/wrir_96-4156/report.html

See also: Hem, J.D., 1985, *Study and interpretation of the chemical characteristics of natural water:* U.S. Geological Survey Water-Supply Paper 2254, 263 p.

See also: Otton, J.K., 1992, *The geology of radon*: U.S. Geological Survey, General Interest Publications of the U.S. Geological Survey, 28 p.

See also: Robillard, P. D., Martin, K. S., and Sharpe, W. E., 1991, *Reducing radon in drinking water*: Pennsylvania State University, Agricultural and Biological Engineering Fact Sheet, SW-135, 4 p.

12 - 16. *Drinking water quality and health*
by P. Kendall1(3/2010)
http://www.ext.colostate.edu/pubs/foodnut/09307.html

Chapter 19 - Concerns - Alcohols & Glycols

As cited, "Alcohol-related intoxications, including methanol, ethylene glycol, diethylene glycol, and propylene glycol, and alcoholic ketoacidosis can present with a high anion gap metabolic acidosis and increased serum osmolal gap, whereas isopropanol intoxication presents with hyperosmolality alone."[1]

With respect to Metabolic acidosis, it is noted as "a clinical disturbance characterized by an increase in plasma acidity. Metabolic acidosis should be considered a sign of an underlying disease process."[2]

As further noted, "there are 3 approaches to understanding acid/base balance"[3]:

Henderson-Hasselbalch approach to acid/base physiology

The Henderson-Hasselbalch equation describes the relationship between blood pH and the components of the H2 CO3 buffering system.

This qualitative description of acid/base physiology allows the metabolic component to be separated from the respiratory components of acid/base balance:

pH = 6.1 + log (HCO3/ H2 CO3) "[4]

Metabolic acidosis can be caused by the following:

• *Increase in the generation of H+ from endogenous (eg, lactate, ketones) or exogenous acids (eg, salicylate, ethylene glycol, methanol)*

• *Inability of the kidneys to excrete the hydrogen from dietary protein intake (type I, IV renal tubular acidosis)*

• *The loss of bicarbonate (HCO3) due to wasting through the kidney (type II renal tubular acidosis) or the gastrointestinal tract (diarrhea)*

• *The kidneys response to a respiratory alkalosis "[5]*

Base excess approach to acid/base physiology

"Unfortunately, the Henderson/Hasselbalch equation is not linear; this nonlinearity of Henderson-Hasselbalch prevents this equation from quantifying the exact amount of bicarbonate deficit in a metabolic acidosis."[6]

"This observation led to the development of a semi quantitative approach, base excess (BE)."[7]

$$BE = (HCO3 - 24.4 + [2.3 \text{ X Hgb} + 7.7] \text{ X } [pH - 7.4]) \text{ X } (1 - 0.023 \text{ X Hgb})"^{8}$$

"Base excess attempts to give a quantitative amount of bicarbonate (mmol) that is required to be added or subtracted to restore 1 L of whole blood to a pH of 7.4 at a pCO2 of 40 mm Hg."[9]

"To standardize BE for hemoglobin, the following formula was developed with improved in vivo accuracy, the standardized base excess (SBE):

$$SBE = 0.9287 \text{ X } (HCO3 - 24.4 + 14.83 \text{ X } [pH - 7.4])."^{10}$$

Strong Ion approach to acid/base physiology

"These classical descriptions of acid/base physiology often failed to account for acid/base findings in critically ill patients."[11]

"An alkalosis was often noted in critically ill patients as their serum albumin level decreased, which could not be quantified by Henderson Hasselbalch or BE."[12]

"Also, the 'dilutional' acidosis frequently encountered after a large infusion of normal saline could not be explained by either of these 2 approaches to acid/base balance."[13]

"Both Henderson Hasselbalch and BE assume that the cations (Ca2+, Mg2+) and anions (Cl-, albumin, PO4-) in plasma remain unchanged in a patient with metabolic acidosis."[14]

"Yet, in critically ill patients, these ions are known to be in dynamic flux."[15]

"During the 1980s, Dr. Peter Stewart developed an acid/base theory (Strong Ion) using quantitative chemistry, which accounted for fluctuations of all the ions dissolved in plasma."[16]

"Based on the requirements for electrical neutrality in any solution as any one of the concentrations of these ions changes, water must dissociate into H+ or OH- to balance the charge."[17]

"The pH in this scheme is not a consequence of the ratio of acid to base in solution but determined by 3 independent variables:

• *Strong ion difference (SID) – Ions almost completely dissociated at physiologic pH.*

$$SID = [Na+ + K+ + Ca2+ + Mg2+] - [Cl- + Lactate-]$$

(Ca2+ and Mg2+ are the concentrations of their ionized forms, Mg2+ X 0.7 = ionized Mg2+ concentration)

• Total weak acid concentration (Atot) – Ions that can exist dissociated (A-) or associated (AH) at physiologic pH (buffers)

Atot = 0.325275 X [albumin] + 2 X [phosphate]

• pCO 2 (mm Hg) "[18]

As such, "the Henderson Hasselbalch equation can be reformulated with variables from the Strong Ion Theory to give a more generalizeable solution to pH.

pH = pK1 ' + log [SID] – Ka – [ATOT]/[Ka + 10–pH]

(K1' is the equilibrium constant for the Henderson-Hasselbalch equation, Ka is the weak acid dissociation constant, and S is the solubility of CO2 in plasma.) '[19]

In addition, "once a metabolic acidosis is suspected by low bicarbonate concentration, an arterial blood gas analysis should be obtained."[20]

"The low HCO3 level can be caused either by a primary metabolic acidosis or as the metabolic compensation for a respiratory alkalosis."[21]

"The direction of the pH will separate metabolic acidosis (pH < 7.35) from a respiratory alkalosis (pH > 7.45)."[22]

As further noted, "Methanol, isopropanol, and propylene glycol are absorbed through normal skin, whereas ethylene glycol and diethylene glycol are absorbed in significant amounts only after the integrity of the skin is breached."[23]

"Inhalation of methanol or topical absorption of ethylene glycol, propylene glycol, isopropanol, and diethylene glycol can produce in-toxications (12,13,30–34), but most intoxications occur after their oral ingestion."[24]

Methanol Intoxication

"Methanol is used in industrial production and is also present in windshield wiper fluid, antifreeze, and model airplane fuel. It is also used in lieu of ethanol. It is colorless and has only a faint odor."[25]

"Methanol intoxication in the United States is uncommon with approximately 1000 to 2000 cases reported each year (approximately 1% of all poisonings)."[26]

"It usually results from accidental ingestion of products containing methanol or ingestion as a method of attempting suicide or is taken in lieu of ethanol when the latter is in short supply."[27]

Ethylene Glycol Intoxication

"Ethylene glycol is used in industrial production and is present in automobile coolants, heat transfer fluids, and runway deicers."[28]

"It is colorless and odorless and has a sweet taste."[29]

"Intoxication has resulted from topical applications of ethylene glycol–containing solutions to burns, but most intoxications result from oral ingestion of ethylene glycol–containing products (most frequently antifreeze), ethanol-based products containing ethylene glycol, or rarely contaminated drinking water."[30]

"Ethylene glycol may be ingested as a cheap substitute for alcohol, taken in an attempt to commit suicide, or ingested accidentally."[31]

"Ethylene glycol intoxication is more frequent than methanol intoxication."[32]

"In 1999, the American Association of Poison Control Centers reported more than 5800 cases reflecting approximately 2% of the poisoning reported in that year in the United States."[33]

"More than 600 of these cases (approximately 11%) occurred in children younger than 6 yr, and more than 800 (approximately 14%) occurred in children aged 6 to 19 yr."[34]

Diethylene Glycol Intoxication

"Diethylene glycol is used in industrial production, is present in brake fluid, and is used as an illegal adulterant in ethanol spirits or in medication."[35]

"Sporadic cases of accidental poisoning or its use as a means of attempting suicide have been described."[36]

"However, this intoxication has primarily been reported in outbreaks in which the diethylene glycol was used as a solvent for medications."[37]

Propylene Glycol

"Propylene glycol is used as a solvent for intravenous, oral, and topical pharmaceutical products and as a major ingredient of some antifreeze and hydraulic fluids."[38]

"Drugs in which it is present include etomidate, phenytoin, diazepam, lorazepam, phenobarbital, nitroglycerin, digoxin, hydralazine, and trimethoprim-sulfamethoxazole."[39]

"Cases of intoxication have been reported after topical administration for treatment for burns and with oral ingestion, but the majority of reported cases have resulted from intravenous administration."[40]

"In this regard, propylene glycol is used as a diluent in benzodiazepines (lorazepam concentration, vol/vol 0.8), which commonly are administered to patients who have seizures, are undergoing alcoholic withdrawal, or are intubated."[41]

"The exact prevalence is not known, but a study of 21 patients receiving benzodiazepines in the intensive care unit revealed that four (19%) had either an increased serum anion gap or a fall in serum bicarbonate concentration consistent with propylene glycol toxicity."[42]

"The serum propylene glycol concentration associated with toxicity has ranged from 12 to 520 mg/dl, but toxicity is most likely to occur at blood concentration in excess of 100 mg/dl."[43]

"Patients with impaired liver or kidney function are said to be at increased risk for developing toxicity."[44]

"In contrast to methanol, ethylene glycol, and diethylene glycol, the mortality of propylene glycol is low, despite its administration to patients with multiple organ dysfunction."[45]

Alcoholic Ketoacidosis

"The syndrome of alcoholic ketoacidosis (AKA) is uncommon in patients with acute ethanol intoxication, being found in <10% of patients."[46]

"It is most frequent in patients who have long-term ethanol intake and liver disease and develop the syndrome after a period of binge drinking and is associated with reduced food intake and episodes of vomiting; the latter might explain the concurrence of metabolic alkalosis noted in some patients."[47]

"A large number of patients with methanol or ethylene glycol intoxication have developed these intoxications after ingesting contaminated ethanol-based products, and, in these cases, serum ethanol levels are high."[48]

"In this regard, increased levels of ketones in the blood or urine have been found in from 2 to 20% of patients with methanol intoxication, consistent with AKA."[49]

"Although AKA often occurs in patients with significant comorbid conditions, the mortality is often low: In a study of 74 patients with AKA, only 1% died)."[50]

<u>Isopropanol</u>

"Isopropanol is used in various industrial products, as a cleaning agent, deicer, and in rubbing alcohol. Isopropanol intoxication results from its accidental ingestion or its use in suicide attempts or when used in lieu of ethanol."[51]

"In 1997, 1999, and 2004, a total of almost 27,000 exposures (most of them minor) were reported to poison control centers; one third were in children who were younger than 6 years."[52]

Mortality is often less than with methanol or ethylene glycol intoxication (0.1% in 2004)."[53]

"Some investigators had suggested that isopropanol concentrations >150 mg/dl produce coma and hypotension and levels >200 mg/dl are said to be incompatible with life."[54]

"By contrast, Lacouture et al. found that severe hypotension with coma was found only with serum isopropanol levels >400 mg/dl."[55]

"Several of the alcohol intoxications can produce severe cellular dysfunction, which can be irreversible, and if they are untreated or treatment is initiated late in their course, then they can be associated with a very high mortality."[56]

"Given the potentially high morbidity and mortality of these intoxications, it is important for the clinician to have a high degree of suspicion for these disorders in cases of high anion gap metabolic acidosis, acute renal failure, or unexplained neurologic disease so that treatment can be initiated early."[57]

Lactic acidosis and diabetic ketoacidosis (DKA) are the most common causes of acute metabolic acidosis."[58]

"Much less frequent but of great importance clinically are the alcohol intoxications; Methanol, ethylene glycol, diethylene glycol, and propylene glycol intoxication and alcoholic ketoacidosis can produce hyperosmolality and metabolic acidosis."[59]

"Isopropanol intoxication is usually associated with hyperosmolality alone."[60]

"Importantly, several of these disorders can be fatal or produce irreversible tissue damage if they are not quickly recognized and treated appropriately."[61]

Footnotes

1. Jeffrey A. Kraut & Ira Kurtz, *Toxic Alcohol Ingestions: Clinical Features,*

Diagnosis, and Management. Clinical Journal of the American Society of Nephrology, January 2008 vol. 3 no. 1 208-225
http://cjasn.asnjournals.org/content/3/1/208.full

1.- 22 Antonia Quinn & Erik D Schraga, *Metabolic Acidosis in Emergency Medicine*
http://emedicine.medscape.com/article/768268-overview

23. Jeffrey A. Kraut & Ira Kurtz, *Toxic Alcohol Ingestions: Clinical Features, Diagnosis, and Management.* Clinical Journal of the American Society of Nephrology, January 2008 vol. 3 no. 1 208-225
http://cjasn.asnjournals.org/content/3/1/208.full

Also see: Barceloux DG, Bond GR, Krenzelok EP, Cooper H, Vale JA: *American Academy of Clinical Toxicology practice guidelines on the treatment of methanol poisoning.* J Toxicol Clin Toxicol40 :415– 446,2002

Also see: Barceloux DG, Krenzelok EP, Olson K, Watson W: *American Academy of Clinical Toxicology practice guidelines on the treatment of ethylene glycol poisoning.* J Toxicol Clin Toxicol37 :537– 560,1999

Also see: Peleg O, Bar-Oz B, Arad I: *Coma in a premature infant associated with the transdermal absorption of propylene glycol.* Acta Paediatr87 :1195– 1196,1998

Also see: Frenia ML, Schauben JL: *Methanol inhalation toxicity.* Ann Emerg Med22 :1919– 1923,1993

Also see: Fligner CL, Jack R, Raisys V, Twiggs GA: *Hyperosmolality induced by propylene glycol: A complication of silver sulfadiazine therapy.* JAMA253 :1606– 1609,1985

Also see: Arditi M, Killner MS: *Coma following use of rubbing alcohol for fever control.* Am J Dis Child141 :237– 238,1987

Also see: Cantarell MC, Fort J, Camps J, Sans M, Piera L, Rodamilans M: *Acute intoxication due to topical application of diethylene glycol.* Ann Intern Med106 :478– 479,1987

24 - 29. Jeffrey A. Kraut & Ira Kurtz, *Toxic Alcohol Ingestions: Clinical Features, Diagnosis, and Management.* Clinical Journal of the American Society of Nephrology, January 2008 vol. 3 no. 1 208-225
http://cjasn.asnjournals.org/content/3/1/208.full

Also see: Lakind JS, McKenna EA, Hubner RP, Tardiff RG: *A review of the comparative mammalian toxicity of ethylene glycol and propylene glycol*. Crit Rev Toxicol29 :331– 365,1999

Also see: Litovitz TL, Klein-Schwartz W, White S, Cobaugh DJ, Youniss J, Drab A, Benson BE*: 1999 annual report of the American Association of Poison Control Centers Toxic Exposure Surveillance System*. Am J Emerg Med18 :517– 574,2000

Also see: Hovda KE, Hunderi OH, Tafjord AB, Dunlop O, Rudberg N, Jacobsen D: *Methanol outbreak in Norway 2002–2004: Epidemiology, clinical features and prognostic signs*. J Intern Med258 :181– 190,2005

Also see: Paasma R, Hovda KE, Tikkerberi A, Jacobsen D: *Methanol mass poisoning in Estonia: Outbreak in 154 patients*. Clin Toxicol45 :152– 157,2007

Also see: Brent J, Lucas M, Kulig K, Rumack BH: *Methanol poisoning in a 6-week-old infant*. J Pediatr118 :644– 646,1991

30 - 37. Jeffrey A. Kraut & Ira Kurtz, *Toxic Alcohol Ingestions: Clinical Features, Diagnosis, and Management*. Clinical Journal of the American Society of Nephrology, January 2008 vol. 3 no. 1 208-225
http://cjasn.asnjournals.org/content/3/1/208.full

Also see: Bruns DE, Herold DA, Rodeheaver GT, Edlich RF: *Polyethylene glycol intoxication in burn patients*. Burns Incl Therm Inj9 :49– 52,1982

Also see: Litovitz TL, Klein-Schwartz W, White S, Cobaugh DJ, Youniss J, Drab A, Benson BE: *1999 annual report of the American Association of Poison Control Centers Toxic Exposure Surveillance System*. Am J Emerg Med18 :517– 574,2000

Also see: Brophy PD, Tenenbein M, Gardner J, Bunchman TE, Smoyer WE: *Childhood diethylene glycol poisoning treated with alcohol dehydrogenase inhibitor fomepizole and hemodialysis*. Am J Kidney Dis35 :958– 962,2000

Also see: Woolf AD: *The Haitian diethylene glycol poisoning tragedy: A dark wood revisited*. JAMA279 :1215– 1216,1998

Also see: Borron SW, Baud FJ, Garnier R: Intravenous 4-methylpyrazole as an antidote for diethylene glycol and triethylene glycol poisoning: A case report. Vet Hum Toxicol39 :26– 28,1997

Also see: Alfred S, Coleman P, Harris D, Wigmore T, Stachowski E, Graudins A: Delayed neurologic sequelae resulting from epidemic diethylene glycol poisoning. Clin Toxicol43 :155– 159,2005

Also see: O'Brien KL, Selanikio JD, Hecdivert C, Placide MF, Louis M, Barr DB, Barr JR, Hospedales CJ, Lewis MJ, Schwartz B, Philen RM, St. Victor S, Espindola J, Needham LL, Denerville K: *Epidemic of pediatric deaths from acute renal failure caused by diethylene glycol poisoning.* JAMA279 :1175– 1180,1998

Also see: Lacey M, Grady D: *A killer in a medicine bottle shakes faith in government.* New York Times October 16,2006

Also see: Bogdanich W, Hooker J: *From China to Panama, a trial of poisoned medicine.* New York Times May 6,2007

38 - 45. Jeffrey A. Kraut & Ira Kurtz, *Toxic Alcohol Ingestions: Clinical Features, Diagnosis, and Management.* Clinical Journal of the American Society of Nephrology, January 2008 vol. 3 no. 1 208-225
http://cjasn.asnjournals.org/content/3/1/208.full

Also see: Peleg O, Bar-Oz B, Arad I: *Coma in a premature infant associated with the transdermal absorption of propylene glycol.* Acta Paediatr87 :1195– 1196,1998

Also see: Brooks DE, Wallace KL: *Acute propylene glycol ingestion.* J Toxicol Clin Toxicol40 :513– 516,2002

Also see: Glover ML, Reed MD: *Propylene glycol: The safe diluent that continues to cause harm.* Pharmacotherapy16 :690– 693,1996

Also see: Arroliga AC, Shehab N, McCarthy K, Gonzales JP: *Relationship of continuous infusion lorazepam to serum propylene glycol concentration in critically ill adults.* Crit Care Med32 :1709– 1714,2004

Also see: Demey HE, Bossaert LL: *Propylene-glycol intoxication and nitroglycerin therapy.* Crit Care Med15 :540 ,1987

Also see: Melvin JJ, Hale PM, Kaye EM: *Iatrogenic propylene glycol intoxication associated with the treatment of status epilepticus.* Ann Neurol44 :569 ,1998

Also see: Arbour R, Esparis B: *Osmolar gap metabolic acidosis in a 60-year-old man treated for hypoxemic respiratory failure: Propylene glycol toxicity caused by escalating lorazepam infusion.* Chest118 :545– 546,2000

Also see: Bekeris L, Baker C, Fenton J, Kimball D, Bermes E: *Propylene-glycol as a cause of an elevated serum osmolality.* Am J Clin Pathol72 :633– 636,1979

Also see: Fligner CL, Jack R, Raisys V, Twiggs GA: *Hyperosmolality induced by propylene glycol: A complication of silver sulfadiazine therapy.* JAMA253 :1606–

1609,1985

Also see: Wilson KC, Reardon C, Theodore AC, Farber HW: *Propylene glycol toxicity: A severe iatrogenic illness in ICU patients receiving IV benzodiazepines—A case series and prospective, observational pilot study.* Chest128 :1674– 1681,2005

Also see: Barnes BJ, Gerst C, Smith JR, Terrell AR, Mullins ME: *Osmol gap as a surrogate marker for serum propylene glycol concentrations in patients receiving lorazepam for sedation.* Pharmacotherapy26 :23– 33,2006

Also see: Doty JD, Sahn SA: *An unusual case of poisoning.* South Med J96 :923– 925,2003

Also see: Parker MG, Fraser GL, Watson DM, Riker RR: *Removal of propylene glycol and correction of increased osmolar gap by hemodialysis in a patient on high dose lorazepam infusion therapy.* Intensive Care Med28 :81– 84,2002

46 - 50. Jeffrey A. Kraut & Ira Kurtz, *Toxic Alcohol Ingestions: Clinical Features, Diagnosis, and Management.* Clinical Journal of the American Society of Nephrology, January 2008 vol. 3 no. 1 208-225
http://cjasn.asnjournals.org/content/3/1/208.full

Also see: Fulop M: *Alcoholic ketoacidosis.* Endocrinol Metab Clin N Am22 :209– 219,1993

Also see: Fulop M, Murthy V, Michilli A, Nalamati J, Qian Q, Saitowitz A: *Serum beta-hydroxybutyrate measurement in patients with uncontrolled diabetes mellitus.* Arch Intern Med159 :381– 384,1999

Also see: Fulop M, Bock J, Ben-Ezra J, Antony M, Danzig J, Gage JS: *Plasma lactate and 3-hydroxybutyrate levels in patients with acute ethanol intoxication.* Am J Med80 :191– 194,1986

Wrenn KD, Slovis CM, Minion GE, Rutkowski R: *The syndrome of alcoholic ketoacidosis.* Am J Med91 :119– 128,1991

Brent J, McMartin K, Phillips S, Aaron C, Kulig K: *Fomepizole for the treatment of methanol poisoning.* N Engl J Med344 :424– 429,2001

Brent J, McMartin K, Phillips S, Burkhart KK, Donovan JW, Wells M, Kulig K: *Fomepizole for the treatment of ethylene glycol poisoning.* N Engl J Med340 :832– 838,1999

Bennett JL, Cary FH, Mitchell GL, Cooper MN: *Acute methyl alcohol poisoning: A review based on experiences in an outbreak of 323 cases.* Medicine32 :431–

463,1953

Megarbane B, Borron SW, Trout H, Hantson P, Jaeger A, Krencker E, Bismuth C, Baud FJ: *Treatment of acute methanol poisoning with fomepizole*. Intensive Care Med27 :1370– 1378,2001

51 - 55. Jeffrey A. Kraut & Ira Kurtz, *Toxic Alcohol Ingestions: Clinical Features, Diagnosis, and Management*. Clinical Journal of the American Society of Nephrology, January 2008 vol. 3 no. 1 208-225
http://cjasn.asnjournals.org/content/3/1/208.full

Also see: Zaman F, Pervez A, Abreo K: *Isopropyl alcohol intoxication: A diagnostic challenge*. Am J Kidney Dis40 :E12 ,2002

Also see: Abramson S, Singh AK: *Treatment of the alcohol intoxications: Ethylene glycol, methanol and isopropanol*. Curr Opin Nephrol Hypertens9 :695– 701,2000

Also see: Litovitz TL, Klein-Schwartz W, White S, Cobaugh DJ, Youniss J, Drab A, Benson BE: *1999 annual report of the American Association of Poison Control Centers Toxic Exposure Surveillance System*. Am J Emerg Med18 :517– 574,2000

Also see: Litovitz TL, Klein-Schwartz W, Dyer KS, Shannon M, Lee S, Powers M: *1997 annual report of the American Association of Poison Control Centers Toxic Exposure Surveillance System*. Am J Emerg Med16 :443– 497,1998

Also see: Watson WA, Litovitz TL, Rodgers GC, Klein-Schwartz W, Reid N, Youniss J, Flanagan A, Wruk KM: *2004 annual report of the American Association of Poison Control Centers Toxic Exposure Surveillance System*. Am J Emerg Med23 :589– 666,2004

Also see: Zaman F, Pervez A, Abreo K: *Isopropyl alcohol intoxication: A diagnostic challenge*. Am J Kidney Dis40 :E12 ,2002

Also see: Chan KM, Wong ET, Matthews WS: *Severe isopropanolemia without acetonemia or clinical manifestations of isopropanol intoxication*. Clin Chem39 :1922– 1925,1993

Also see: Lacouture PG, Wason S, Abrams A, Lovejoy FH: *Acute isopropyl alcohol intoxication: diagnosis and management*. Am J Med75 :680– 686,1983

56 - 61. Jeffrey A. Kraut & Ira Kurtz, *Toxic Alcohol Ingestions: Clinical Features, Diagnosis, and Management*. Clinical Journal of the American Society of Nephrology, January 2008 vol. 3 no. 1 208-225
http://cjasn.asnjournals.org/content/3/1/208.full

Also see: Gabow PA, Kaehny WD, Fennessey PV, Goodman SI, Gross PA, Schrier RW: *Diagnostic importance of an increased serum anion gap.* N Engl J Med303 :854– 858,1980

Kraut JA, Madias NE: *Serum anion gap: Its uses and limitations in clinical medicine.* Clin J Am Soc Nephrol2 :162– 174,2007

Hojer J: *Severe metabolic acidosis in the alcoholic: Differential diagnosis and management.* Hum Exp Toxicol15 :482– 488,1996

Zaman F, Pervez A, Abreo K: *Isopropyl alcohol intoxication: A diagnostic challenge.* Am J Kidney Dis40 :E12 ,2002

Abramson S, Singh AK: *Treatment of the alcohol intoxications: Ethylene glycol, methanol and isopropanol.* Curr Opin Nephrol Hypertens9 :695– 701,2000

Jacobsen D, McMartin KE: *Methanol and ethylene-glycol poisonings: Mechanism of toxicity, clinical course, diagnosis and treatment.* Med Toxicol Adverse Drug Exp1 :309– 334,1986

Hovda KE, Hunderi OH, Tafjord AB, Dunlop O, Rudberg N, Jacobsen D: *Methanol outbreak in Norway 2002–2004: Epidemiology, clinical features and prognostic signs.* J Intern Med258 :181– 190,2005

Karlsonstiber C, Persson H: *Ethylene-glycol poisoning: Experiences from an epidemic in Sweden.* J Toxicol Clin Toxicol30 :565– 574,1992

Hanif M, Mobarak MR, Ronan A, Rahman D, Donovan JJ, Bennish ML: *Fatal renal failure caused by diethylene glycol in paracetamol elixir*: The Bangladesh epidemic. BMJ311 :88– 91,1995

Ferrari LA, Giannuzzi L: *Clinical parameters, postmortem analysis and estimation of lethal dose in victims of a massive intoxication with diethylene glycol.* Forensic Sci Int153 :45– 51,2005

Jacobsen D, Ovrebo S, Ostborg J, Sejersted OM: *Glycolate causes the acidosis in ethylene glycol poisoning and is effectively removed by hemodialysis.* Acta Med Scand216 :409– 416,1984

Kerns W, Tomaszewski C, McMartin K, Ford M, Brent J: *Formate kinetics in methanol poisoning.* J Toxicol Clin Toxicol40 :137– 143,2002

Chapter 20 - Concerns - Chlorinated Aromatics

With respect to bleaching, "chlorine dioxide and chlorine -- because of their fundamentally different chemistries -- react in distinct ways with organic compounds, and as a result generate very different byproducts."[1]

"It is this difference that explains the superior environmental performance of chlorine dioxide in paper making and scrubbers."[2]

"Technically speaking, both chlorine and chlorine dioxide are oxidizing agents -- electron receivers."[3]

"Chlorine has the capacity to take in two electrons, whereas chlorine dioxide can absorb five."[4]

"This property, along with the complex, but well known, ways chlorine combines with lignin (the cellular adhesive in wood tissue), explains the basic difference between the two compounds."[5]

"In the chlorine-based bleaching process, about 10 percent of the chlorine combines directly with lignin which has 'aromatic' components."[6]

"Aromatic compounds have atoms arranged in rings, and they may have other atoms, such as chlorine, attached to these rings."[7]

"Within the group of chlorinated aromatics, which can be toxic to some organisms, are the infamous dioxins."[8]

"Chlorinated aromatic compounds are largely considered as undesirable synthetic pollutants."[9]

"Adsorbable organic halogen (AOX) measurements are commonly used as a group parameter to monitor pollution in the environment."[10]

"Industries worldwide are required to comply with stringent environmental standards for chloroaromatic substances."[11]

"However, industry is not necessarily the only source of these "xenobiotic" compounds."[12]

"Recently, high AOX levels which greatly exceed diffuse anthropogenic deposition have been reported to occur in pristine forest environments."[13]

"Worldwide, the AOX concentrations encountered ranged from 5 to 350 mg of Cl- per kg (dry weight)."[14]

"A natural origin is plausible, since several microorganisms are known to form de novo up to 1,500 different organohalogen metabolites."[15]

"Most wood- and forest litter-degrading fungi are basidiomycetes, and several strains from many different genera have been reported to produce de novo chlorinated aromatic compounds."[16]

"Chlorinated anisyl metabolites (CAM) are biosynthesized by Lepista, Stropharia, and Bjerkandera spp."[17]

"Drosophilin A (tetrachloromethoxyphenol) and derivatives are produced by Psathyrella, Fomes, Mycena, and Phellinus spp."[18]

"Also the production of other chloroaromatics, such as chloroan-thraquinones by Dermocybe spp.[19], strobilurin B by Strobilurus spp.[20], mycenon by Mycena spp.[21], and 3-chloro-4-hydroxyphenylacetate by Marasmius spp.[22], has been reported."[23]

Footnotes

1 - 8. *Chloride Dioxide*
http://home.windstream.net/mikeric/Odor/clo2.htm

9 - 23. E. De Jong, Jim A. Field, Henri-Eric Spinner, Joannes B. P. A. Wijnberg & Jan A. De Bont, *Significant Biogenesis of Chlorinated Aromatics by Fungi in Natural Environments*. Applied & Environmental Microbiology, Vol. 60, No. 1, Jan. 1994, p. 264-270
http://docs.google.com/viewer?a=v&q=cache:_bpkIBJ2mJMJ:www.ncbi.nlm.nih.
gov/pmc/articles/PMC201298/pdf/aem00018-0286.pdf+Chlorinated+Aromatics+-
+Organics&hl=en&gl=ca&pid=bl&srcid=ADGEESjKQ5KpRBM82CVFt5sA-
T_CKLWP3FBotNSB5m8UFXDQfdR9tUa5RKcyG_wSJl0jd9oix1vnfeq0KdnPn
TatTWfxSbp11fDN1v6lWYFmpORW70Mz8ohhlEUGh1mLleCFVMUmIEvj&si
g=AHIEtbT5xDLzxnIi_2obRB5rRL2VJFH0rA

Also see: Asplund, G., and A. Grimvall. 1991. *Organohalogens in nature-more widespread than previously assumed*. Environ. Sci. Technol. 25:1346-1350.

Also see: Asplund, G., and A. Grimvall. 1991. *Organohalogens in nature-more widespread than previously assumed*. Environ. Sci. Technol. 25:1346-1350.

Also see: Gribble, G. W. 1992. *Naturally occurring organohalogen compounds-a survey*. J. Nat. Prod. (Lloydia) 55:1353-1395.

Also see: Neidleman, S. L., and J. Geigert. 1986. *Biohalogenation: principles, basic roles and applications.* Ellis Horwood Ltd., Chichester, England.

Also see: Siuda, J. F., and J. F. DeBernardis. 1973. *Naturally occurring halogenated organic compounds.* Lloydia (Cinci.) 36:107-143.

Also see: de Jong, E., J. A. Field, J. A. F. M. Dings, J. B. P. A. Wijnberg, and J. A. M. de Bont. 1992. *De novo biosynthesis of chlorinated aromatics by the white-rot fungus Bjerkandera sp. BOS55: formation of 3-chloro-anisaldehyde from glucose.* FEBS Lett. 305:220-224.

Also see: Pfefferle, W., H. Anke, M. Bross, and W. Steglich. 1990. *Inhibition of solubilized chitin synthase by chlorinated aromatic compounds isolated from mushroom cultures.* Agric. Biol. Chem. 54:1381-1384.

Also see: Thaller, V., and J. L. Turner. 1972. *Natural acetylenes. Part XXXV. Polyacetylenic acid and benzenoid metabolites from cultures of fungus Lepista diemii Singer.* J. Chem. Soc. Perkins Trans. 1972:2032-2034.

Also see: Anchel, M. 1952. *Identification of drosophilin A as p-methoxytetrachlorophenol.* J. Am. Chem. Soc. 74:2943.

Also see: Butruille, D., and X. A. Dominguez. 1972. *Un nouveau produit naturel: dimethoxy-1,4 nitro-2 trichloro-3,5,6 benzene.* Tetrahedron Lett. 1972:211-212.

Also see: Hsu, C. S., M. Suzuji, and Y. Yamada. 1971. *Chemical constituents of fungi. I. 1,4-dimethoxy-2,3,5,6-tetrachlorobenzene (0-methyldrosophilin A) from Phellinus yucatensis.* Chem. Abstr. 75:115864a.

Also see: Singh, P., and S. Rangaswami. 1966. *Occurence of o-methyldrosophilin A in Fomes fastuosus Lev.* Tetrahedron Lett. 1966:1229-1231.

Also see: van EUk, G. W. 1975. *Drosophilin A methyl ether from Mycena megaspora.* Phytochemistry 14:2506.

Also see: Gruber, I. 1970. *Anthraquinone pigments of the genus Dermocybe and their use in taxonomic evaluation.* Z. Pilzkd. 36:95-112.

Also see: Anke, T., G. Schramm, B. Schwalge, B. Steffan, and W. Steglich. 1984. *Antibiotika aus Basidiomyceten, XX: synthese von Stro-*

*bilurin A und Revision der Stereochemie der naturlichen Stro-
bilurine*. Liebigs Ann. Chem. 1616-1625.

Also see: Hautzel, R., H. Anke, and W. S. Sheldrick. 1990. *Mycenon, a
new metabolite from a Mycena species TA 87202 (Basidiomycetes) as an inhibitor of isocitrate-lyase*. J. Antibiot. 43:1240-

Also see: Haggblom, M. M. 1992. *Microbial breakdown of halogenated
aromatic pesticides and related compounds*. FEMS Microbiol.
Rev. 103:29-72.

Chapter 21 - Concerns - Haloacetic acids

As noted, "Haloacetic acids (HAAs) are a group of compounds that can form when the chlorine used to disinfect drinking water reacts with naturally occurring organic matter (e.g., decaying leaves and vegetation)."[1]

"Haloacetic acids (HAAs) are a type of chlorination disinfection by-product (CDBP) that are formed when the chlorine used to disinfect drinking water reacts with naturally occurring organic matter (NOM) in water."[2]

"Haloacetic acids are a relatively new disinfection by-product."[3]

"HAAs are a collection of several different compounds."[4]

"The sum of Bromodichloroacetic Acid (BrCl2AA), Dibromochloroacetic Acid (Br2ClAA), and Tribromoacetic Acid (Br3AA) concentrations is known as HAA3."[5]

"The sum of Monochloroacetic Acid (ClAA), Monobromoacetic Acid (BrAA), Dichloroacetic Acid (Cl2AA), Trichloroacetic Acid (Cl3AA), and Dibromoacetic Acid (Br2AA) concentrations are known as HAA5."[6]

"HAA6 refers to the sum of HAA5 and Bromochloroacetic Acid (BrClAA) concentrations. HAA6 and HAA3 together make up HAA9."[7]

"The reported HAAs value refer to the sum of the concentration of six haloacetic acid compounds which include mono-, di-, and trichloroacetic acids, and mono- and dibromoacetic acids, and bromochloroacetic acid."[8]

"The Guidelines for Canadian Drinking Water Quality (GCDWQ) recommend a maximum acceptable concentration (MAC) of 80 micrograms per litre (µg/L) for HAAs in drinking water, based on a locational running annual average of a minimum of quarterly samples taken in the distribution system."[9]

"Changes in body, liver, kidney and testes weights were observed in studies with rats."[10]

"A health-based target concentration of 0.1 mg/L can be calculated for MCA in drinking water."[11]

Footnotes

1. *Guidelines for Canadian Drinking Water Quality: Guideline Technical Document - Haloacetic Acids*
http://www.hc-sc.gc.ca/ewh-semt/pubs/water-eau/haloaceti/summary-sommaire-

eng.php

2 - 9. *Haloacetic acids (HAAs)*
http://www.env.gov.nl.ca/env/waterres/quality/drinkingwater/haa.html

10 - 11. *Guidelines for Canadian Drinking Water Quality: Guideline Technical Document - Haloacetic Acids*
http://www.hc-sc.gc.ca/ewh-semt/pubs/water-eau/haloaceti/summary-sommaire-eng.php

Chapter 22 - Concerns - Dioxins & Furans

As cited, "Polychlorinated dibenzo-para-dioxins (dioxins) and polychlorinated dibenzofurans (furans) are two groups of planar tricyclic compounds that have very similar chemical structures and properties."[1]

"They may contain between 1 and 8 chlorine atoms; dioxins have 75 possible positional isomers and furans have 135 positional isomers. They are generally very insoluble in water, are lipophilic and are very persistent."[2]

"Neither dioxins nor furans are produced commercially, and they have no known use. They are by-products resulting from the production of other chemicals."[3]

"Dioxins may be released into the environment through the production of pesticides and other chlorinated substances."[4]

"Furans are a major contaminant of PCBs (Polychlorinated biphenyls)."[5]

"Both dioxins and furans are related to a variety of incineration reactions, and the synthesis and use of a variety of chemical products."[6]

"Dioxins and furans have been detected in emissions from the incineration of hospital waste, municipal waste, hazardous waste, car emissions, and the incineration of coal, peat and wood."[7]

This site also includes, "Dioxin is formed by burning chlorine-based chemical compounds with hydrocarbons."[8]

"The major source of dioxin in the environment comes from waste-burning incinerators of various sorts and also from backyard burn-barrels."[9]

"Dioxin pollution is also affiliated with paper mills which use chlorine bleaching in their process and with the production of Polyvinyl Chloride (PVC) plastics and with the production of certain chlorinated chemicals (like many pesticides)."[10]

Continuing, "of the 210 dioxins and furans, 17 contribute most significantly to the toxicity of complex mixtures."[11]

"Dioxins and furans are some of the most toxic chemicals known to science."[12]

"A draft report released for public comment in September 1994 by the US Environmental Protection Agency clearly describes dioxin as a serious public health threat."[13]

"The public health impact of dioxin may rival the impact that DDT had on public health in the 1960's. According to the EPA report, not only does there appear to be no 'safe' level of exposure to dioxin, but levels of dioxin and dioxin-like chemicals have been found in the general US population that are 'at or near levels associated with adverse health effects'."[14]

In answer to the question, Does dioxin cause cancer?

This site also responds with a resounding "Yes."

"The EPA report confirmed that dioxin is a cancer hazard to people."

"In 1997, the International Agency for Research on Cancer (IARC) -- part of the World Health Organization -- published their research into dioxins and furans and announced on February 14, 1997, that the most potent dioxin, 2,3,7,8-TCDD, is a now considered a Group 1 carcinogen, meaning a 'known human carcinogen'."[15]

Also, in January 2001, the U.S. National Toxicology Program upgraded 2,3,7,8-TCDD from "Reasonably Anticipated to be a Human Carcinogen" to "Known to be a Human Carcinogen."[16]

For example, "a bioassay of a mixture of 1,2,3,6,7,8- and 1,2,3,7,8,9-hexachlorodibenzo-p-dioxin (HCDD) for possible carcinogenicity was conducted by administering the test material by gavage to Osborne-Mendel rats and B6C3F1 mice for 104 weeks."[17]

"Fifty rats and 50 mice of each sex were administered HCDD suspended in a vehicle of 9:1 corn oil-acetate 2 days per week for 104 weeks at doses of 1.25, 2.5, or 5 µg/kg/wk for rats and male mice and 2.5, 5, or 10 µg/kg/wk for female mice."[18]

"Seventy-five rats and 75 mice of each sex served as vehicle controls."[19]

"In addition, one untreated control group containing 25 rats and 25 mice of each sex was present in the HCDD treatment room, and one untreated control group containing 25 rats and 25 mice of each sex was present in the vehicle control room. All surviving animals were killed at 105 to 108 weeks."[20]

"In male rats, hepatocellular carcinomas or neoplastic nodules occurred at low incidences that were dose related (P=0.003)."[21]

"In a direct comparison, the incidence of these tumors in the high-dose group was higher (P=0.022) than that in the corresponding vehicle-control groups."[22]

"In female rats, hepatocellular carcinomas, adenomas, or neoplastic nodules occurred at incidences that were dose related (P<0.001), and in direct comparisons the incidences of these tumors in the mid-and high-dosed groups were significantly higher (P=0.006 and P<0.001, respectively) than those in the corresponding vehicle-control group."[23]

"In male mice, hepatocellular carcinomas or adenomas occurred at incidences that were dose related (P=0.001), and in a direct comparison the incidence of these tumors in the high-dose group was significantly higher (P=0.001) than that in the corresponding vehicle-control group."[24]

"In female mice, hepatocellular carcinomas or adenomas occurred at incidences that were dose-related (P=0.002), and the incidence of these tumors in the high-dose group was significantly higher (P=0.004) than that in the corresponding vehicle-control group."[25]

"Complex nonneoplastic toxic liver lesions were seen in all dosed groups of rats and mice."[26]

"Compound-associated hyperplastic lesions of the lung were also found in both male and female rats."[27]

As also reported, "a 2003 re-analysis of the cancer risk from dioxin reaffirmed that there is no known 'safe dose' or 'threshold' below which dioxin will not cause cancer."[28]

"A July 2002 study shows dioxin to be related to increased incidence of breast cancer."[29]

This site concluded, "Dioxin/furans cause cancer, immunotoxicity and reproductive disorders in humans even at low levels of toxicity."[30]

"They are endocrine disrupters that damage the liver, cause miscarriages, birth deformities and weakening of the immune system."[31]

"They concentrate in breast tissue affecting nursing infants."[32]

"EPA says they are 300,000 times more carcinogenic than DDT."[33]

Footnotes

1 - 7. *Dioxins and Furans*
http://www.popstoolkit.com/about/chemical/dioxin.aspx

8 - 10. *Dioxin Homepage*
http://www.ejnet.org/dioxin/

11. *Dioxins and Furans*

http://www.popstoolkit.com/about/chemical/dioxin.aspx

12 - 16. *Dioxin Homepage*
http://www.ejnet.org/dioxin/

16 - 27. *Bioassay of a Mixture of 1,2,3,6,7,8-Hexachlorodibenzo-p-dioxin and 1,2,3,7,8,9-Hexachlorodibenzo-p-dioxin (Gavage) for Possible Carcinogenicity (CAS No. 57653-85-7,CAS No. 19408-74-3)*
http://ntp.niehs.nih.gov/?objectid=0705D188-ED98-D7A5-9C560C37DB9F973A

28 - 29. *Dioxin Homepage*
http://www.ejnet.org/dioxin/

30. *Montana River Action: Toxins*
http://www.montanariveraction.org/toxins.html

Chapter 23 - Concerns - Glyphosate

As reported, "Glyphosate is an organic solid of odorless white crystals."[1]

"Glyphosate is a non-selective herbicide used on many food and non-food crops as well as non-crop areas such as roadsides."[2]

"When applied at lower rates, it serves as a plant growth regulator."[3]

"The most common uses include control of broadleaf weeds and grasses in: hay/pasture, soybeans, field corn; ornamentals, lawns, turf, forest plantings, greenhouses, rights-of-way."[4]

As further cited, "in a paper published in the European Journal of Agronomy in October 2009, Huber and co-author G.S. Johal, from Purdue's department of botany and plant pathology, state that the widespread use of glyphosate that we see today in agriculture in the United States can 'significantly increase the severity of various plant diseases, impair plant defense to pathogens and diseases, and immobilize soil and plant nutrients rendering them unavailable for plant use'."[5]

"Further, the authors state that glyphosate stimulates the growth of fungi and enhances the virulence of pathogens such as Fusarium and 'can have serious consequences for sustainable production of a wide range of susceptible crops'."[6]

As well, in 1974, Congress passed the Safe Drinking Water Act, with a corresponding maximum contaminant level goals (MCLG) for contaminants – "any physical, chemical, biological or radiological substances or matter in water."[7]

"The MCLG for glyphosate is 0.7 mg/L or 700 ppb."[8]

Footnotes

1 - 4. *Basic Information about Glyphosate in Drinking Water*
http://water.epa.gov/drink/contaminants/basicinformation/glyphosate.cfm

5 - 6. *Scientist warns of dire consequences with widespread use of glyphosate*
http://www.non-gmoreport.com/articles/may10/consequenceso_widespread_glyphosate_use.php

7 - 8. *Basic Information about Glyphosate in Drinking Water*
http://water.epa.gov/drink/contaminants/basicinformation/glyphosate.cfm

Chapter 24 - Concerns - Chlorite

As noted, "the occurrence of toxic chemicals in drinking water supplies has been on the rise over the past few decades, primarily as a result of increased manufacturing, soil erosion and agriculture."[1]

"Many heavy metals and pesticides have been monitored for years, but concern is growing about the increase in outbreak and contamination due to the rise in nationwide occurrences."[2]

"The U.S. Environmental Protection Agency implements and oversees the Safe Drinking Water Act, which includes a list of contaminants that must be screened for in all public drinking water supplies."[3]

"These contaminants include microorganisms, disinfectants and their byproducts, radionuclides and various organic and inorganic chemicals."[4]

"Microorganisms include things like cryptosporidium, fecal coliforms, viruses and legionella and come from poorly treated human and animal fecal waste."[5]

"Chemicals used to disinfect drinking water are also hazardous to human health and include things like bromate and chlorite."[6]

"Inorganic chemicals include heavy metals like cadmium, copper and lead, and other compounds such as nitrate from fertilizer and asbestos."[7]

"The majority of organic compounds include chemicals such as glyphosate and atrazine, common ingredients in many agricultural pesticides."[8]

As further reported, "Chlorite is a general name for several minerals that are difficult to distinguish by ordinary methods. These minerals are all a part of the Chlorite Group of minerals"[9] [see **Appendix 18a**, appearing below].

"The chlorites are often, but not always considered a subset of the larger silicate group, The clays"[10] [see **Appendix 18b**, appearing below].

"The general formula for chlorite is $(Fe, Mg, Al)_6(Si, Al)_4O_{10}(OH)_8$."[11]

"However there are several different minerals that are a part of the chlorite group of minerals."[12]

"The above formula is only a generalization of the more common members of this group."[13]

"For practical reasons most of the chlorites will be considered here as a single mineral, chlorite."[14]

"Chlorites are generally green and crystallize in the monoclinic symmetry system."[15]

"They all have a basal cleavage due to their stacked structure."[16]

"Chlorites typically form flaky microscopic crystals and it is this reason that they are sometimes included in the clay group of minerals."[17]

"However chlorites also form large individual tabular to platy crystals that are unlike most of the other clay minerals."[18]

Footnotes

1 - 8. *Toxic Water Pollutants*
By John Marton, Jun 8, 2010
http://www.livestrong.com/article/142654-toxic-water-pollutants/

9 - 18. *The Mineral Chlorite*
http://www.galleries.com/Chlorite

Appendix 18a

The Chlorite Group

"The Chlorite Group This group is not always considered a part of the clays and is sometimes left alone as a separate group within the phyllosilicates."[1]

"It is a relatively large and common group although its members are not well known. These are some of the recognized members"[2]:

"•*Amesite*
• *(Mg, Fe)4Al4Si2O10(OH)8*

•*Baileychlore*
• *(Zn, Fe+2, Al, Mg)6(Al, Si)4O10(O, OH)8*

•*Chamosite*
• *(Fe, Mg)3Fe3AlSi3O10(OH)8*

•*Clinochlore (kaemmererite)*
• *(Fe, Mg)3Fe3AlSi3O10(OH)8*

•*Cookeite*
• *LiAl5Si3O10(OH)8*

•*Corundophilite*
• *(Mg, Fe, Al)6(Al, Si)4O10(OH)8*

•*Daphnite*
• *(Fe, Mg)3(Fe, Al)3(Al, Si)4O10(OH)8*

•*Delessite*
• *(Mg, Fe+2, Fe+3, Al)6(Al, Si)4O10(O, OH)8*

•*Gonyerite*
• *(Mn, Mg)5(Fe+3)2Si3O10(OH)8*

•*Nimite*
• *(Ni, Mg, Fe, Al)6AlSi3O10(OH)8*

•*Odinite*
• *(Al, Fe+2, Fe+3, Mg)5(Al, Si)4O10(O, OH)8*

•*Orthochamosite*
• *(Fe+2, Mg, Fe+3)5Al2Si3O10(O, OH)8*

•Penninite
• *(Mg, Fe, Al)6(Al, Si)4O10(OH)8*

•Pannantite
• *(Mn, Al)6(Al, Si)4O10(OH)8*

•Rhipidolite (prochlore)
• *(Mg, Fe, Al)6(Al, Si)4O10(OH)8*

•Sudoite
• *(Mg, Fe, Al)4 - 5(Al, Si)4O10(OH)8*

•Thuringite
• *(Fe+2, Fe+3, Mg)6(Al, Si)4O10(O, OH)8"* [3]

"The term chlorite is used to denote any member of this group when differentiation between the different members is not possible. The general formula is X4-6Y4O10(OH, O)8."[4]

"The X represents one or more of aluminum, iron, lithium, magnesium, manganese, nickel, zinc or rarely chromium."[5]

"The Y represents either aluminum, silicon, boron or iron but mostly aluminum and silicon."[6]

"The gibbsite layers of the other clay groups are replaced in the chlorites by a similar layer that is analogous to the oxide brucite."[7]

"The structure of this group is composed of silicate layers sandwiching a brucite or brucite-like layer in between, in an s-b-s stacking sequence similar to the above groups."[8]

"However, in the chlorites, there is an extra weakly bonded brucite layer in between the s-b-s sandwiches."[9]

"This gives the structure an s-b-s b s-b-s b sequence. The variable amounts of water molecules would lie between the s-b-s sandwiches and the brucite layers."[10]

Reference to:

1 – 10. *The Clay Minerals*
http://www.galleries.com/Clays_Group

Appendix 18b

The Clay Minerals

"The Clay Minerals are a part of a general but important group within the phyllosilicates that contain large percentages of water trapped between their silicate sheets."[1]

"Most clays are chemically and structurally analogous to other phyllosilicates but contain varying amounts of water and allow more substitution of their cations."[2]

"There are many important uses and considerations of clay minerals."[3]

"They are used in manufacturing, drilling, construction and paper production."[4]

"They have great importance to crop production as clays are a significant component of soils."[5]

"It is the physical characteristics of clays (more so than chemical and structural characteristics) that define this group"[6]:

"•*Clay minerals tend to form microscopic to sub microscopic crystals.*

•*They can absorb water or lose water from simple humidity changes.*

•*When mixed with limited amounts of water, clays become plastic and are able to be molded and formed in ways that most people are familiar with as children's clay.*

•*When water is absorbed, clays will often expand as the water fills the spaces between the stacked silicate layers.*

•*Due to the absorption of water, the specific gravity of clays is highly variable and is lowered with increased water content.*

•*The hardness of clays is difficult to determine due to the microscopic nature of the crystals, but actual hardness is usually between 2 - 3 and many clays give a hardness of 1 in field tests.*

•*Clays tend to form from weathering and secondary sedimentary processes with only a few examples of clays forming in primary igneous or metamorphic environments.*

•*Clays are rarely found separately and are usually mixed not only with other clays but with microscopic crystals of carbonates, feldspars, micas and quartz.*"[7]

"Clay minerals are divided into four major groups."[8]

These are the important clay mineral groups:[9]

<u>The Kaolinite Group</u>

"This group has three members (kaolinite, dickite and nacrite) and a formula of $Al_2Si_2O_5(OH)_4$."[10]

"The different minerals are polymorphs, meaning that they have the same chemistry but different structures (polymorph = many forms)."[11]

"The general structure of the kaolinite group is composed of silicate sheets (Si_2O_5) bonded to aluminum oxide/hydroxide layers $(Al_2(OH)_4)$ called gibbsite layers."[12]

"The silicate and gibbsite layers are tightly bonded together with only weak bonding existing between the s-g paired layers."[13]

"Uses: In ceramics, as a filler for paint, rubber and plastics and the largest use is in the paper industry that uses kaolinite to produce a glossy paper such as is used in most magazines."[14]

<u>The Montmorillonite/Smectite Group</u>

"This group is composed of several minerals including pyrophyllite, talc, vermiculite, sauconite, saponite, nontronite and montmorillonite."[15]

"They differ mostly in chemical content. The general formula is $(Ca, Na, H)(Al, Mg, Fe, Zn)_2(Si, Al)_4O_{10}(OH)_2 - xH_2O$, where x represents the variable amount of water that members of this group could contain."[16]

"Talc's formula, for example, is $Mg_3Si_4O_{10}(OH)_2$."[17]

"The gibbsite layers of the kaolinite group can be replaced in this group by a similar layer that is analogous to the oxide brucite, $(Mg_2(OH)_4)$."[18]

"The structure of this group is composed of silicate layers sandwiching a gibbsite (or brucite) layer in between, in an s-g-s stacking sequence."[19]

"The variable amounts of water molecules would lie between the s-g-s sandwiches."[20]

"Uses: Are many and include a facial powder (talc), filler for paints and rubbers, an electrical, heat and acid resistant porcelain, in drilling muds and as a plasticizer in molding sands and other materials."[21]

The Illite (or The Clay-mica) Group

"This group is basically a hydrated microscopic muscovite."[22]

"The mineral illite is the only common mineral represented, however it is a significant rock forming mineral being a main component of shales and other argillaceous rocks."[23]

"The general formula is (K, H)Al2(Si, Al)4O10(OH)2 - xH2O, where x represents the variable amount of water that this group could contain."[24]

"The structure of this group is similar to the montmorillonite group with silicate layers sandwiching a gibbsite-like layer in between, in an s-g-s stacking sequence."[25]

"The variable amounts of water molecules would lie between the s-g-s sandwiches as well as the potassium ions."[26]

"Uses: A common constituent in shales and is used as a filler and in some drilling muds."[27]

The Chlorite Group appears above in **Appendix 18a**.

Reference to:

1 – 27. *The Clay Minerals*
http://www.galleries.com/Clays_Group

Chapter 25 - More concerns drinking water

Cyanide – Inorganic

As cited, "a cyanide is a chemical compound that contains the cyano group, -C≡N, which consists of a carbon atom triple-bonded to a nitrogen atom."[1]

"Cyanides most commonly refer to salts of the anion CN−."[2]

"Most cyanides are highly toxic."[3]

"In organic chemistry compounds containing a -C≡N group are known as nitriles and compounds that contain the -N≡C group are known as isocyanides."[4]

Nitriles

"Compounds having the structure RC≡N; thus C-substituted derivatives of hydrocyanic acid, HC≡N. In systematic nomenclature, the suffix nitrile denotes the triply bound ≡N atom, not the carbon atom attached to it."[5]

Isocyanides

"The isomerHN+≡C−of hydrocyanic acid, HC≡N, and its hydrocarbyl derivatives RNC (RN+≡C−)."[6]

"Organic nitriles and isocyanides are far less toxic because they do not release cyanide ions easily."[7]

As further reported, "Cyanide is produced naturally in the environment by various bacteria, algae, fungi and numerous species of plants including beans (coffee, chickpeas and lima), fruits (seeds and pits of apple, cherry, pear, apricot, peach and plum), almond and cashew nuts, vegetables of the cabbage family, grains (alfalfa, and sorghum), roots (cassava, potato, radish and turnip), white clover and young bamboo shoots."[8]

"Incomplete combustion during forest fires is believed to be a major environmental source of cyanide, and incomplete combustion of articles containing nylon produces cyanide through depolymerization."[9] [see **Appendix 19**, appearing below].

Also, "Cyanide is produced in the human body and exhaled in extremely low concentrations with each breath. It is also produced by over 1,000 plant species including sorghum, bamboo and cassava. Relatively low concentrations of cyanide can be highly toxic to people and wildlife."[10]

"Cyanide is acutely toxic to humans. Liquid or gaseous hydrogen cyanide and alkali salts of cyanide can enter the body through inhalation, ingestion or absorption through the eyes and skin. The rate of skin absorption is enhanced when the skin is cut, abraded or moist; inhaled salts of cyanide are readily dissolved and absorbed upon contact with moist mucous membranes."[11]

"The toxicity of hydrogen cyanide to humans is dependent on the nature of the exposure. Due to the variability of dose-response effects between individuals, the toxicity of a substance is typically expressed as the concentration or dose that is lethal to 50% of the exposed population (LC50 or LD50)."[12]

"The LC50 for gaseous hydrogen cyanide is 100-300 parts per million. Inhalation of cyanide in this range results in death within 10-60 minutes, with death coming more quickly as the concentration increases."[13]

"Inhalation of 2,000 parts per million hydrogen cyanide causes death within one minute."[14]

"The LD50 for ingestion is 50-200 milligrams, or 1-3 milligrams per kilogram of body weight, calculated as hydrogen cyanide."[15]

"For contact with unabraded skin, the LD50 is 100 milligrams (as hydrogen cyanide) per kilogram of body weight."[16]

In addition, "n 1974, Congress passed the Safe Drinking Water Act. This law requires EPA to determine safe levels of chemicals in drinking water which do or may cause health problems. These non-enforceable levels, based solely on possible health risks and exposure, are called Maximum Contaminant Level Goals."[17]

"The MCLG for cyanide has been set at 0.2 parts per million (ppm) because EPA believes this level of protection would not cause any of the potential health problems described below"[18]:

Short-term:

"EPA has found cyanide to potentially cause the following health effects when people are exposed to it at levels above the MCL for relatively short periods of time: rapid breathing, tremors and other neurological effects."[19]

Long-term:

"Cyanide has the potential to cause the following effects from a lifetime exposure at levels above the MCL: weight loss, thyroid effects, nerve damage."[20]

As further reported:

Effects on Wildlife:

"Although cyanide reacts readily in the environment and degrades or forms complexes and salts of varying stabilities, it is toxic to many living organisms at very low concentrations."[21]

Aquatic Organisms:

"Fish and aquatic invertebrates are particularly sensitive to cyanide exposure. Concentrations of free cyanide in the aquatic environment ranging from 5.0 to 7.2 micrograms per liter reduce swimming performance and inhibit reproduction in many species of fish."[22]

"Other adverse effects include delayed mortality, pathology, susceptibility to predation, disrupted respiration, osmoregulatory disturbances and altered growth patterns."[23]

"Concentrations of 20 to 76 micrograms per liter free cyanide cause the death of many species, and concentrations in excess of 200 micrograms per liter are rapidly toxic to most species of fish."[24]

"Invertebrates experience adverse nonlethal effects at 18 to 43 micrograms per liter free cyanide, and lethal effects at 30 to 100 micrograms per liter (although concentrations in the range of 3 to 7 micrograms per liter caused death in the amphipod Gammarus pulex)."[25]

Birds:

"Reported oral LD50 for birds range from 0.8 milligrams per kilogram of body weight (American racing pigeon) to 11.1 milligrams per kilogram of body weight (domestic chickens)."[26]

"Symptoms including panting, eye blinking, salivation and lethargy appear within one-half to five minutes after ingestion in more sensitive species, and up to ten minutes after ingestion by more resistant species."[27]

"Exposures to high doses resulted in deep, labored breathing followed by gasping and shallow intermittent breathing in all species. Mortality typically occurred in 15 to 30 minutes; however birds that survived for one hour frequently recovered, possibly due to the rapid metabolism of cyanide to thiocyanate and its subsequent excretion."[28]

"Ingestion of WAD cyanide solutions by birds my cause delayed mortality. It appears that birds may drink water containing WAD cyanide that is not immediately fatal, but which breaks down in the acidic conditions in the stomach

and produces sufficiently high cyanide concentrations to be toxic.[29]

Mammals:

"Cyanide toxicity to mammals is relatively common due to the large number of cyanogenic forage plants such as sorghum, sudan grasses and corn. Concentrations of cyanide in these plants are typically highest in the spring during blooming. Dry growing conditions enhance the accumulation of cyanogenic glycosides in certain plants as well as increase the use of these plants as forage."[30]

"Reported oral LD50 for mammals range from 2.1 milligrams per kilogram of body weight (coyote) to 6.0-10.0 milligrams per kilogram of body weight (laboratory white rats). Symptoms of acute poisoning usually occur within ten minutes of ingestion, including: initial excitability with muscle tremors; salivation; lacrimation; defecation; urination; labored breathing; followed by muscular incoordination, gasping and convulsions."[31]

"In general, cyanide sensitivity for common livestock decreases from cattle to sheep to horses to pigs; deer and elk appear to be relatively resistant."[32]

Footnotes

1 - 4. *Cyanide*
http://en.wikipedia.org/wiki/Cyanide

Also see: *IUPAC Gold Book cyanides*
http://goldbook.iupac.org/C01486.html
"Salts and C-organyl derivatives of hydrogen cyanide, HC≡N, e.g. CH3C≡N methyl cyanide (acetonitrile), NaCN sodium cyanide, PhC(=O)CN benzoyl cyanide."

Also see: Greenwood, N. N.; & Earnshaw, A. (1997). *Chemistry of the Elements (2nd Edn.)*, Oxford:Butterworth-Heinemann. ISBN 0-7506-3365-4.

Also see: G. L. Miessler and D. A. Tarr *"Inorganic Chemistry"* 3rd Ed, Pearson/Prentice Hall publisher, ISBN 0-13-035471-6.

Also see: *"Environmental and Health Effects of Cyanide"*. International Cyanide Management Institute. 2006. Retrieved 4 August 2009.
http://www.cyanidecode.org/cyanide_environmental.php

5 - 6. *IUPAC Gold Book cyanides*
http://goldbook.iupac.org/C01486.html

7. *Cyanide*

http://en.wikipedia.org/wiki/Cyanide

8 - 16. *"Environmental and Health Effects of Cyanide"*. International Cyanide Management Institute. 2006. Retrieved 4 August 2009.
http://www.cyanidecode.org/cyanide_environmental.php

17 - 20. *Drinking Water Contaminants- Cyanide*
http://www.freedrinkingwater.com/water-contamination/cyanide-contaminants-removal-water.htm

21 - 32. *"Environmental and Health Effects of Cyanide"*. International Cyanide Management Institute. 2006. Retrieved 4 August 2009.
http://www.cyanidecode.org/cyanide_environmental.php

Appendix 19

Environmental and Health Effects

Reference to: Environmental and Health Effects of Cyanide
http://www.cyanidecode.org/cyanide_environmental.php

As cited, "once released in the environment, the reactivity of cyanide provides numerous pathways for its degradation and attenuation":

Complexation:

"Cyanide forms ionic complexes of varying stability with many metals. Most cyanide complexes are much less toxic than cyanide, but weak acid dissociable complexes such as those of copper and zinc are relatively unstable and will release cyanide back to the environment. Iron cyanide complexes are of particular importance due to the abundance of iron typically available in soils and the extreme stability of this complex under most environmental conditions. However, iron cyanides are subject to photochemical decomposition and will release cyanide if exposed to ultraviolet light. Metal cyanide complexes are also subject to other reactions that reduce cyanide concentrations in the environment, as described below."

Precipitation:

"Iron cyanide complexes form insoluble precipitates with iron, copper, nickel, manganese, lead, zinc, cadmium, tin and silver. Iron cyanide forms precipitates with iron, copper, magnesium, cadmium and zinc over a pH range of 2-11."

Adsorption:

"Cyanide and cyanide-metal complexes are adsorbed on organic and inorganic constituents in soil, including oxides of aluminum, iron and manganese, certain types of clays, feldspars and organic carbon. Although the strength of cyanide retention on inorganic materials is unclear, cyanide is strongly bound to organic matter."

Cyanate:

"Oxidation of cyanide to less toxic cyanate normally requires a strong oxidizing agent such as ozone, hydrogen peroxide or hypochlorite. However, adsorption of cyanide on both organic and inorganic materials in the soil appears to promote its oxidation under natural conditions."

Thiocyanate:

"Cyanide reacts with some sulfur species to form less toxic thiocyanate. Potential sulfur sources include free sulfur and sulfide minerals such as chalcopyrite ($CuFeS_2$), chalcocite (Cu_2S) and pyrrhotite (FeS), as well as their oxidation products, such as polysulfides and thiosulfate."

Volatilization:

"At the pH typical of environmental systems, free cyanide will be predominately in the form of hydrogen cyanide, with gaseous hydrogen cyanide evolving slowly over time. The amount of cyanide lost through this pathway increases with decreasing pH, increased aeration of solution and with increasing temperature. Cyanide is also lost through volatilization from soil surfaces."

Biodegradation:

"Under aerobic conditions, microbial activity can degrade cyanide to ammonia, which then oxidizes to nitrate. This process has been shown effective with cyanide concentrations of up to 200 parts per million. Although biological degradation also occurs under anaerobic conditions, cyanide concentrations greater than 2 parts per million are toxic to these microorganisms."

Hydrolysis:

"Hydrogen cyanide can be hydrolyzed to formic acid or ammonium formate. Although this reaction is not rapid, it may be of significance in ground water where anaerobic conditions exist."

Other work by these authors

2003 Category: Non-Fiction
Re-introduction of gray wolves to some natural habitats to counter the effects of over-populated wild herd animals due to the elimination of their main predator
Mallenby, Patricia
ISBN: 0973281324
Canadiana: 20030165210
LC Call no.: QL737 C22 M325 2003 fol.
Dewey: 599.773 13
AMICUS No. 28839674

2003 Category: Non-Fiction
Genetics - Gregor Johann Mendel: small sampling problems?
Mallenby, Patricia & Jeremy
ISBN: 0973281359
Canadiana: 20040126161
LC Class no.: QH430
Dewey: 576.5 14
AMICUS No. 29768606

2003 Category: Non-Fiction
Have chocolate manufacturers capitalized on recent research about the health benefits of chocolate to increase their chocolate sales?
Mallenby, Jeremy
ISBN: 0973281316
Canadiana: 20040054942
LC Call no.: HD9200 A2 M34 2003 fol.
Dewey: 338.4/7664153 22
AMICUS No. 28839676

2007 Category: Non-Fiction
Essays in World History: An Undergraduate Perspective
Mallenby, Patricia & Jeremy
ISBN: 978-0-9780593-1-6
Number of pages: 425

2008 Category: Non-Fiction
So You Want To Be A Probation Officer? A Review of Some Supreme Court and Provincial Court Cases to Help Clarify the Nature of Probation Orders and the Work of Probation Officers
Mallenby, Patricia & Jeremy

ISBN: 9781897518779
Number of pages: 407
2008 Category: Non-Fiction
Is He Our Sister? Was She Our Father? Expert Medical &
Scientific Evidence Re-define Identity, Marriage, Family &
Children: A Chronological Review of Court Decisions &
Legislative Accommodation
Mallenby, Patricia & Jeremy
ISBN: 9781926626604
Number of pages: 186

2010 Category: Non-Fiction
Aboriginal Self Government & Other Self Determination Issues
ISBN-10: 1602646406, ISBN-13: 978-1602646407
Number of pages: 334

2010 Category: Non-Fiction
Rewarding Probationer Compliance
ISBN-10: 1453862013, ISBN-13: 978-1453862018
Number of pages: 372

www.ingramcontent.com/pod-product-compliance
Lightning Source LLC
Chambersburg PA
CBHW081437170526
45166CB00008B/2226